The Chorleywood Bread Process

Related titles:

Using cereal science and technology for the benefit of consumers:
Proceedings of the 12th International ICC Cereal and Bread Congress,
2004
(ISBN-13: 978-1-85573-961-1; ISBN-10: 1-85573-961-5)
These proceedings from one of the leading conferences on the subject
summarise some of the most significant recent developments in
breadmaking and cereals processing.

Bread making: improving quality
(ISBN-13: 978-1-85573-553-8; ISBN-10: 1-85573-553-9)
Edited by one of the leading authorities on breadmaking, and with an
international team of contributors, this volume reviews the key research
on how ingredients and processes in breadmaking determine bread
quality.

Baking problems solved
(ISBN-13: 978-1-85573-564-4; ISBN-10: 1-85573-564-4)
Based on decades of experience, this highly practical book provides the
answers to key technical problems experienced by bakers.

Details of these books and a complete list of Woodhead titles can be
obtained by:

• visiting our web site at www.woodheadpublishing.com
• contacting Customer Services (e-mail: sales@woodhead-
  publishing.com; fax: +44 (0) 1223 893694; tel.: +44 (0) 1223 891358 ext.
  30; address: Woodhead Publishing Limited, Abington Hall, Abington,
  Cambridge CB1 6AH, England)

If you would like to receive information on forthcoming titles in this
area, please send your address details to: Francis Dodds (address, tel. and
fax as above; e-mail: francisd@woodhead-publishing.com). Please confirm
which subject areas you are interested in.

The Chorleywood Bread Process

Stanley P. Cauvain and Linda S. Young

BakeTran

CRC Press
Boca Raton Boston New York Washington, DC

WOODHEAD PUBLISHING LIMITED
Cambridge England

Published by Woodhead Publishing Limited, Abington Hall, Abington,
Cambridge CB1 6AH, England
www.woodheadpublishing.com

Published in North America by CRC Press LLC, 6000 Broken Sound Parkway, NW,
Suite 300, Boca Raton, FL 33487, USA

First published 2006, Woodhead Publishing Limited and CRC Press LLC
© 2006, Woodhead Publishing Limited
The authors have asserted their moral rights.

British Library Cataloguing in Publication Data
A catalogue record for this book is available from the British Library.

Library of Congress Cataloging in Publication Data
A catalog record for this book is available from the Library of Congress.

Woodhead Publishing ISBN-13: 978-1-85573-962-8 (book)
Woodhead Publishing ISBN-10: 1-85573-962-3 (book)
Woodhead Publishing ISBN-13: 978-1-84569-143-1 (e-book)
Woodhead Publishing ISBN-10: 1-84569-143-1 (e-book)
CRC Press ISBN-10: 0-8493-9131-8
CRC Press order number: WP9131

The publishers' policy is to use permanent paper from mills that operate a
sustainable forestry policy, and which has been manufactured from pulp which is processed
using acid-free and elementary chlorine-free practices. Furthermore, the publishers ensure
that the text paper and cover board used have met acceptable environmental accreditation
standards.

Typeset by SNP Best-set Typesetter Ltd., Hong Kong
Printed by TJ International Ltd, Padstow, Cornwall, England

Contents

9 Variety bread production with the Chorleywood

Preface

In the UK the setting up of 'research associations' in 1919 as an industry–government partnership was an attempt to stimulate research and to create stronger links between research and the application of knowledge arising from it. The concept was based on subscription income raised by particular industry sectors which would be matched pound for pound from the public purse. It was with that concept in mind that the British Baking Industries Research Association (BBIRA) was established shortly after the Second World War. Its chosen location for its research laboratories was a commuter suburb of London, Chorleywood in Metro Land.

Much has been written and said about the processes involved in innovation and invention. Most new inventions come from the incremental increase in knowledge about a given subject through patient research and study of the subject concerned. Questions which drive researchers include 'why does that happen?' and 'how does that happen?'. Sometimes when focused thought combines with the results of methodical study and mixes with a little bit of serendipity those 'Eureka! moments' occur.

In many academic environments of the past the link between research and applications was not always highest on the priority list for sponsored research. In a research association, curiosity-driven research has to be combined with commercial needs and so other questions are quickly added to the list. Questions such as 'can I make the product more effectively?', 'can I make a product of better quality?' and 'can I develop a product that no one else has to give myself a commercial edge?' quickly rise up the list of priorities for the researchers involved.

Baking has always been, and remains, a complex mixture of science, technology, craft and artistry and because of this the study of the subject has always required a blend of different skills and personalities. As history has shown, given the right blend of skills, personalities and patient, thorough research Eureka! moments are sure to follow. Such a blend of skills and individuals arrived in BBIRA, Chorleywood, in the late 1950s and early 1960s and they were to have their own Eureka! moments, some of which resulted in the development of what became known as the Chorleywood Bread Process (CBP) – the subject of this book.

The introduction of the CBP was a watershed moment for the UK and soon after for the international baking world. It came at a time when certainly the UK baking industry was in the process of dramatic change. Its impact went beyond that of any other breadmaking process, mixing machine or set of magic ingredients. The ripples that the stone which was the CBP spread after being thrown into the 'pond' which is baking travelled around the globe and are still spreading today. Changes in improver and ingredient technology, process and equipment design, and even the nature of commercial operations sparked by the launch of the CBP form the basis of the modern baking industry.

This is not to say that everything that we work with today in baking is a direct result of the CBP. The success of the CBP was because it disturbed the equilibrium of the baking industry and tilted the future in favour of the science and technology of the product rather than the craft and the artistry. In doing so, it prompted a greater questioning of the then state of knowledge about baking. As the underpinning tenets of baking were more closely questioned and scrutinised, so the value of existing knowledge was better appreciated and systemised and the status of baking as a scientific and technical subject was enhanced.

We both joined the Flour Milling and Baking Research Association (FMBRA) when it was based at Chorleywood and were fortunate to be able to work in various capacities with some of the giants of that organisation. The launch of the CBP placed the Research Association firmly on the international cereals stage but never led to complacency. New knowledge was always being uncovered, new discoveries made and new innovations exploited. The CBP itself evolved and adapted as new research offered new opportunities.

In some ways the story of the CBP is a story of an organisation and the people who worked there. We are proud to have worked there and are pleased to help to tell some of the story that makes the CBP unique in the annals of baking.

Stanley P. Cauvain
25 years with FMBRA in various roles and 10 years as Director of
Cereals and Cereal Processing Division, Campden & Chorleywood Food
Research Association

Linda S. Young
16 years at FMBRA in various roles and 10 years as Technology
Transfer Manager in the Cereals and Cereal Processing Division,
Campden & Chorleywood Food Research Association

Acknowledgements

There have been many contributors to the story of the CBP. Some of the key players have been named in different parts of this book. No doubt there are many, many more who made contributions but who have not been named; this is not to belittle their individual contributions but rather to accept that the list would be very long. We are pleased to acknowledge that many of the illustrations used come from the FMBRA, Chorleywood, archive. Figures 7.1–7.4 and 9.1–9.3 were supplied by BakeTran.

1

The basic principles of breadmaking

The aim of all breadmaking processes is essentially the same, namely to convert wheat flour and other ingredients into a light, aerated and palatable food. Bread is probably the oldest 'processed' food. Its discovery probably took place in the Middle East where the origins of cereal farming also lie in antiquity. Early forms of bread would have been very different from how we see it in industrialised countries today and it would probably be closest in character to the modern flat breads of the Middle East. The move to improve the digestibility of the wild grass seed forerunners of early wheat types through fermentation and baking represents a major step in the evolution of human food production. The discovery that dough left for long periods of time would increase in volume before being subjected to the high temperatures of baking identified the basis of fermentation (gas production). There is no doubt that the changes in the rheological character of the dough would have been observed by those in charge of food production. The combined effect of these changes is for the subsequent baked mass to increase in volume during baking and to give a product with an even softer, more digestible character and different flavour.

There are a few basic steps which form the basis of all modern breadmaking. They are as follows:

1 The mixing of wheat flour and water, together with yeast and salt, and other specified ingredients in appropriate ratios.
2 The development of a gluten (hydrated protein) structure in the dough by the transfer of energy during mixing, commonly referred to as 'kneading'.
3 The incorporation of air bubbles into the dough during mixing.

4   The modification of the dough's rheological properties during knead-
     ing which improves its ability to expand when gas pressures in the
     dough increase from the generation of carbon dioxide gas in the fer-
     menting dough. This stage of dough development may also be referred
     to as 'ripening' or 'maturing'.
5   The creation and modification of particular flavours in the dough
     through fermentation, especially when prolonged bulk resting times
     are used.
6   A subdivision of the bulk dough mass into unit pieces for subsequent
     processing.
7   For some products, a preliminary modification of the shape of the
     divided dough pieces.
8   A short delay between the dividing–initial shaping and the final
     shaping to modify dough piece rheology for final shaping.
9   A final shaping of the dough pieces to achieve their required form.
10   The fermentation and expansion of the shaped dough pieces during
     'proof'.
11   Further expansion of the dough pieces and fixation of the final bread
     structure during baking.

There are two underpinning processes to consider in breadmaking, both of
which are required to create a light and aerated palatable food:

1   Gas production, which refers to the generation of carbon dioxide gas as
     a natural consequence of yeast fermentation. Provided that the yeast
     cells in the dough remain viable (alive) and sufficient substrate (food)
     for the yeast is available, then gas production will continue.
2   Gas retention, which refers to the ability of the dough to retain the
     carbon dioxide gas produced by yeast fermentation and to expand in
     proving and baking.

Fermentation occurs whenever yeast is present and the conditions are
favourable for yeast activity. The extent of bakers' yeast activity in a bread
dough depends on several factors such as the length of time, the dough temperature, the dough pH and the presence of any inhibitors (natural or
added). Bakery terminology can be confusing to both specialist and non-
specialist alike. This is particularly true for the processing terms which are
used to distinguish the stages of carbon dioxide generation by the yeast.
Carbon dioxide gas generation starts quickly in most bread doughs and may
occur to a significant extent in some cases during the mixing process itself
(especially if a long mixing time is involved). However, most bakers consider that fermentation does not occur until after mixing has finished.
   If the dough is left to ferment in bulk after mixing, then the process is
generally referred to as 'bulk fermentation' or 'floortime'. During this time
the bulk volume of the dough increases as it expands through the generation of carbon dioxide gas. A key function of the bulk fermentation process

is to modify the rheological properties of the gluten in the dough and thereby to improve the quality of the final bread product. The length of time required to achieve the required changes varies with the qualities of the flour used in the formulation (e.g. protein content), the yeast level and dough temperature. Part of the way through the bulk fermentation time the dough may be remixed ('knocked back' or 'punched') in order to modify dough rheology and control fermentation further. The whole of the bulk fermentation process is commonly referred to as 'dough development' or 'dough maturation' and is an integral part of the production of yeast-fermented products.

Dough development is a relatively undefined term used to describe the complex changes which are set in motion when the ingredients are first mixed. These are changes associated with the formation of gluten, which requires both the hydration of the proteins in the flour and the application of energy through the process of mixing or kneading. However, there is more to dough development than a simple kneading process. The process of developing the gluten structure of a bread dough improves its ability to retain the carbon dioxide gas generated by yeast fermentation in the stages which follow mixing and before the yeast is killed in the oven. This improvement is particularly important when the dough pieces reach the oven because it is in the early stages of baking before the dough has set that yeast activity is at its greatest and large quantities of carbon dioxide gas are generated in the dough. If the dough pieces are to continue to expand at this time, then the dough must be able to retain a large quantity of this gas and it can only do this if a gluten structure with the most appropriate rheological properties has been created (Cauvain, 1998a).

There is an intimate relationship between the rheological properties of dough and dough development. Rheology may be briefly described as the study of how materials flow and behave when stresses and strains are applied. Dough is described as a viscoelastic material, i.e. it displays both viscous and elastic properties; however, since it is a non-Newtonian material its rheological properties are complicated and a number of different techniques have been evolved to measure such properties (Faridi and Faubion, 1989). The rheological or handling properties of dough are determined by many factors, a few of which are as follows:

1   The properties of the proteins in wheat flour.
2   The temperature of the dough.
3   The length of any times for which the dough rests.
4   The addition of functional ingredients, such as oxidising agents.
5   The impact of 'work' on the dough.

The rheological properties of a given dough vary in a given breadmaking process depending on where in the process it is being assessed. For example, at the end of a resting period (e.g. the bulk fermentation period) the dough will be softer (i.e. easier to mould) than it was when it first left the mixer.

This difference in dough behaviour is important and the transition in dough-handling properties which take place while the dough rests in bulk has been classically described as dough development or maturing. Whatever the term used, the purpose is to indicate that the rheological properties of the bulk dough are optimal for the process of converting it into unit pieces (dividing) for shaping and moulding. This does not mean that the rheological properties of the dough are optimal for all subsequent stages of processing nor does it mean that the only manner in which the rheological properties of dough can be optimised is through the use of a bulk resting stage.

Defining an optimum dough rheology is difficult, not least because of the lack of a definitive way in which to measure it. This arises because dough is non-Newtonian and so, when sufficient force is applied to it, e.g. a high shear as may happen during dough-moulding stages, irreversible changes to the dough properties occur. In bakers' terms this may be seen as damage or tearing of the dough surface which may contribute to quality defects in the baked products (see Chapter 7). The importance of understanding and appreciating dough rheology cannot be underestimated in the mechanical processing stages. When dough is moulded by hand, the skilled baker can recognise subtle variations in dough rheology and adjust moulding pressures accordingly; no mechanical moulder currently possesses the same degree of subtle touch when it comes to handling dough.

While it may be difficult to offer a precise definition for bread dough rheology, it is helpful to consider in general terms what properties are required and their relevance. These are as follows:

1   *Resistance to deformation.* This is may be equated to dough softness and in general terms the dough should be have as little resistance to deformation as possible. This means that it is easy to change the shape of the dough during moulding and thereby to limit damage to the relatively delicate dough structures.
2   *Elasticity.* This refers to the degree to which a dough piece will attempt to regain its original shape after the deforming force has been removed. In practical terms the dough should have a minimum of elasticity to avoid defects with product shape because of 'spring-back' after moulding. If spring-back does occur, then a common practice is to increase moulding pressures, which in turn leads to more damage to dough structures.
3   *Extensibility.* This is related to the ability of the dough to stretch as the gas pressures from yeast fermentation build up. In general, dough should have good extensibility (and minimal elasticity) to yield significant dough expansion during proving and baking.
4   *Stickiness.* This is a most undesirable property for dough since it is the common cause of problems in moulding. In practice the distinction between softness and stickiness is blurred. This is because the most

common way in which bakers compensate for dough stickiness is by removing water from the recipe which makes the dough 'stiffer' and easier to work. The reality is that the excessive mechanical working of dough is often the cause of dough stickiness and, if this is limited, then relatively high dough water levels can be used.

The creation of dough with a more extensible gluten character is especially important for improved gas retention while reductions in dough resistance and elasticity play major roles in the modification of bubble structures during processing. Until about 40 years ago the necessary rheological changes were achieved by mixing the dough and allowing the dough to rest in bulk for a defined period of time (several hours) before dividing and processing of the unit pieces and 'bulk fermentation' processes are still used today.

The concept of using 'energy' to 'develop' dough has been around for some time. For example Ford (1975) described how competition breads in the period from 1910 to 1930 were 'developed' by passing short-time bulk fermentation doughs through pastry brakes which imparted low levels of energy to the dough accumulated through successive passages. As we now know, some of the improvements seen in product cell structure at that time arose because of the truncation of the gas bubble populations in the dough. The large gas bubbles in the fermented dough were burst and the surviving gas bubble population consisted of the smaller and more uniform bubbles which contributed to the finer and more uniform cell structure in the final product.

Swanson and Working (1926) published the first paper entitled 'Mechanical modification of dough'. They reported how mixing small doughs for longer periods or at higher speeds produced better bread than short mixing (3 min) and 100 min of bulk fermentation. This work was largely ignored, probably because it was only carried out on a small scale, until the late 1950s and early 1960s when the increasing industrialisation of bakeries led to intensive research and resulted in the breadmaking processes of today. One of the most significant developments from that intensive research was the invention of the Chorleywood Bread Process (CBP) and it is this process which is discussed in detail in this book.

2

A brief history of the Chorleywood Bread Process

The origins of the CBP lie with research work carried out in the late 1950s and early 1960s by the British Baking Industries Research Association (BBIRA) based at Chorleywood, Hertfordshire, UK. The BBIRA was one of many research associations which were active in the UK at that time. The basis of its formation in 1946 was a government–industry partnership in which the value of the subscription income raised by an industry sector (in this case, baking) was matched with equal funding from the UK government. The intention was to carry out 'generic' research on behalf of the baking industry which would then have equal access to the output of that research to apply within individual member companies.

The UK baking industry of the 1950s was very different from that seen today. At that time the BBIRA was active in studying the fundamentals of dough mixing which would ultimately lead to the development of the CBP bread and other fermented products that were made in a large number of relatively small family-owned bakery businesses. Today, the UK baking industry is dominated by a small number of large industrial plant bakeries and in-store bakeries in supermarkets with relatively few family-sized 'traditional' bakery businesses.

The 1950s were a time of great change for the UK baking industry. Typically bulk fermentation for 2 or 3 h was used to make bread at that time. The Weston Company considered that they needed to import flour direct from Canada to make the best bread. This action provoked millers into protecting their sales of flour by buying up many independent plant bakeries and so the UK milling and baking groups that we know today were born.

It was also a time of great experiment in the UK baking industry. There was much interest in the continuous mixing of bread doughs which was seen as a natural progression for the plant bakery, although in most cases with some fermentation after mixing. Continuous mixers came to the UK in large numbers. For example, from the USA came the Wallace and Tiernan Do-Maker (1957), the Amflow and the Oakes special bread process (Williams, 1975). The lack of sustainable success of these processes in the UK was associated with the uniform structure and weak crumb structure that they produced. American bread characteristics were in sharp contrast with the leaner formula and denser UK bread products of the time and they found little favour with UK consumers.

The early work of BBIRA concentrated on ingredient and equipment testing and providing services and consultancy to baker members. One area of study during their early years was that of dough mixing. Indeed *BBIRA Report 42* (Collins *et al.*, 1959) provides data on their early investigations into the possibilities of using mechanical dough development using a brew system and continuous development to replace bulk fermentation. BBIRA had many members, mainly small and independent bakers but in the later 1950s the nature of the BBIRA client base began to change as companies became larger and started to build their own research facilities.

Into this changing picture came Dr George Elton as the newly appointed Director of Research at BBIRA in 1958. He quickly recognised that he needed to change the pattern of work at the BBIRA if he was to retain their members. To achieve this, Elton identified six research areas for BBIRA activity and one of these was continuous mixing (Elton and Devlin, 1959). He chose T.H. (Bill) Collins as the bakery technologist to work on continuous mixing. In the early days, Bill worked alone on the task but quickly he began to organise and increase his work rate by 'acquiring' more help. Elton recognised the value of the work that was coming from Bill and quickly added to his team. In 1959, Dr Norman Chamberlain had joined BBIRA as Scientific Liaison Officer but very soon Elton was to ask Chamberlain to join with Collins to lead a task force concentrating on dough mixing and bread making.

At first the team continued to work with continuous mixers carrying out two or three experiments a day. Even in those early days the critical nature of work input was appreciated when they discovered that the quality of bread could be improved by taking a continuous mixed dough and then giving it further mixing in a laboratory Morton Z-blade mixer (Fig. 2.1). The duration of the second mixing step appeared to be critical and Elton made the crucial observation that the quantity of work might be important. Soon they found that they could obtain the same quantity of work from the laboratory Z-blade mixer as they could from the continuous mixer (Fig. 2.2). Ever eager to increase the number of experiments that they could perform in a day the team switched their attention to the small-scale mixer and improved their experimental output by an order of magnitude.

With around 30 mixings a day they made rapid progress towards establishing the key features of the mixing and development processes. In

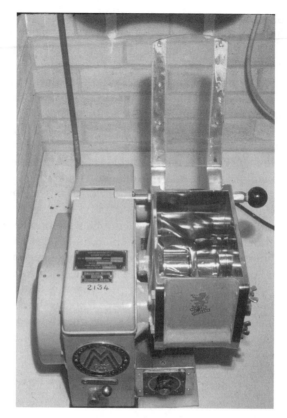

Fig. 2.1   Laboratory-scale Z-blade mixer fitted with timer.

keeping with the attitudes of the time they published their work in a series of ground-breaking *BBIRA Reports* (see Further reading for details). In some ways the team would come to regret the openness with information because it later prevented them from patenting the process that they had developed and from deriving direct financial benefit for the BBIRA. However, it should be appreciated that the BBIRA had an obligation to its members to inform them how the research monies were being spent. To some extent the BBIRA were obliged to disseminate freely the results of their work to their members.

During their work on dough mixing, the BBIRA team was to show that they could make bread from controlled-energy batch mixing without bulk fermentation which was identical with or superior in quality (as defined by bread volume, softness and cell structure) to that they made with low-speed mixing and bulk fermentation. They could not say the same for the comparison between continuously mixed doughs and bulk fermented doughs. While working with a variety of mixers the BBIRA team found that,

Fig. 2.2 Laboratory-scale Z-blade mixer in the BBIRA bread bakery fitted with a watt–hour meter.

although the crumb cell structure varied from one mixer to another, the optimum work input (in terms of bread volume) remained essentially the same (the relationship between cell structure and bread volume and the contribution of different mixing actions is now more fully understood; see Chapter 5). The BBIRA team calculated energy levels in the dough and found that the optimum for the range of flours available at that period of time was 0.4 hp min/lb (Collins *et al.*, 1960), equivalent to 5 W h/lb, 11 W h/kg or 42 kJ/kg of dough in the mixer (Fig. 2.3).

The BBIRA team knew they were on to a new process but needed to scale up to convince bakers of its commercial value. They had a number of discussions with existing bread mixer manufacturers without being able to

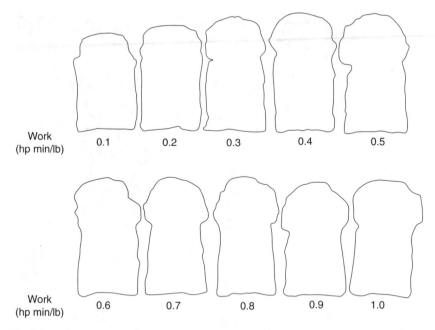

Fig. 2.3　Effect of work in the CBP: upper line, left to right, from 0.1 to 0.5 hp min/lb; bottom line, left to right, from 0.6 to 1.0 hp min/lb, in steps of 0.1 hp min/lb.

raise significant interest from any of them who remained steadfast to existing low-speed batch and continuous mixers. This was a time of frustration for all concerned but then serendipity intervened. A former colleague of Norman Chamberlain who worked for J. Sainsbury asked Chamberlain if he would like to see the high-speed mixer that they used for making paste for pies. It performed the job in 20 s. Norman Chamberlain and Bill Collins immediately went to see the machine and quickly recognised its potential for delivering the required energy at the required rate. They borrowed a mixing machine on the basis that, if it failed to match their requirements, they would return it.

The mixer arrived at the Chorleywood site from the Tweedy factory in Lancashire complete with green painted lid and galvanised bowl. With the mixer in place at Chorleywood the BBIRA team set about dough making, checked temperature rises during mixing and realised that the machine did indeed have the potential for delivering the energy that they required in the time that they required. The resident physicist at Chorleywood, John Cornford, obtained and wired up a watt–hour meter and the team began to prove their small-scale findings. Once proven, the team needed a better-looking mixer and within 6 weeks with the help of the engineering skills of Ken Pickles a stainless steel version was produced to the BBIRA specification.

In early 1961 the team began to identify the key features of this new bread-making system and agonised over what to call it. Eventually they settled on a name and in July 1961 the BBIRA team launched the Chorley-wood Breadmaking Process named after the town in which BBIRA was located. The launch of the new process was heralded with the publication of BBIRA Report 59 (Chamberlain *et al.*, 1961).

From 1961 the pace of development increased as the BBIRA team set off with the mixer in a van to demonstrate this new process to bakers around the UK. Even by today's standard the schedule was tough. The team would arrive at a new location and would have to solve all sorts of problems related to working space and power, and to work with all sorts of ovens, but they managed well. All in all, the team demonstrated to over 1000 bakers in a little over 12 months on their whirlwind tour of the British Isles. They only demonstrated to members of the BBIRA at this time. If you turned up and were not a member of BBIRA, you would be requested to join before you entered the bakery to see the demonstration. Extracts taken from the personal log of Bill Collins from 1961 given in Table 2.1 shows just how busy the Chorleywood team was in 1961.

An important part in commercial acceptance of the CBP was played by the then Chief Executive Officer of the Rank Hovis group, Lord Rank. He visited the BBIRA in the early days and the team were to show off proudly their new bread process but, as they did so, the watt–hour meter that they were using caught fire and filled with acrid-smelling smoke which leaked out into the bakery. It clearly did not deter Lord Rank from recognising the potential of the CBP and on the day he is reported to have said (Collins, personal communication), 'Ah, there's nothing like the smell of fresh bread!'

The pilot mixer underwent change as Ken Pickles of Tweedy tried his first experiments with the application of vacuum in order to provide a less open cell structure in the final bread. He was actually building on the Wallace and Tiernan Do-Maker requirement to remove excess air from the mixing-chamber headspace to avoid mixing too much air into the dough, causing the structure of the bread to become too open. After he had collapsed several mixing bowls trying to obtain higher and higher vacuum levels, Pickles settled for what he could readily achieve in practice, about half an atmosphere (0.5 bar absolute but then expressed as 15 in mercury) (Pickles, 1968). Some years later, Dr Norman Chamberlain (1983) was to say, 'It is only in retrospect that the result seems surprising and the explanation obscure, making this an illustration of the importance of not letting theory stand in the way of practical trial and error.' Considerable interest in the early days was to come from the Rank Hovis bread subsidiary British Bakeries. Working for them on continuous mixing at the time was Willy Anderson, who was to play a significant role in the acceptance of the CBP with the first commercial Tweedy mixer using partial vacuum going into production in August 1962.

Table 2.1 Extract from the personal log of Bill Collins for 1961

Date	Activity
14 August	Demonstration of CBP on Tweedy to representatives of ABF [Associated British Foods]
15 August	Demonstration of CBP to Badex representatives
16 August	Demonstration of CBP to British Bakeries representatives
17 August	Visit to Hales for CBP work on BP [Baker Perkins] high-speed mixer
21 August	Test bakes at Hales
23 August	Test bakes at Hales
25 August	Test bakes at Hales
30 August	[Bill] Pringle [British Arkady] visits RA [BBIRA]
8 September	Hales visit
12 September	CBP demonstration to CWS [Cooperative Wholesale Society]
13 September	Tweedy's baker visits RA
19 September	BSP [British Soya Products] visits RA
20 September	Visit to BP with CBP bread samples
26 September	Demonstration at Liverpool Cooperative
27 September	Demonstration at Warburtons. Tweedy (Burnley) in evening to see first Tweedy mixer for bread.
28 September	Tweedy am, Arkady pm, Simon-AMF evening.
2 October	Lord Rank visits RA for CBP demonstration
3 October	To Hales bakery
5 October	Demonstration of CBP to London judges
6 October	Japanese visit RA
9 October	Groups from Holland visit RA
10 October	Hales
11 October	Kent, Jones and Amos [Laboratories] to see demonstration of CBP
13 October	Simpson of Admiralty to see CBP
25 October	Bread for [BBIRA] Annual General Meeting display

(Extract and adaptation courtesy of T.H. Collins.)

The commercial uptake of the CBP was strongly influenced by the improver companies. At its introduction the CBP used an oxidising agent in addition to the bromate then added to flours at the mill. There had been significant amounts of experimental work using potassium iodate but it was not permitted as a bread improver in the UK. Instead ascorbic acid was chosen and added as a water-based solution (Fig. 2.4). Also to be added was fat with a suitable slip point. The development of the powder-based improver allowed the necessary oxidising agents to be added as part of a powder and this hastened the acceptance of the CBP, even though at the time both improver and fat had to be added manually at the start of the mixing cycle. Improver and ingredient companies benefited from the introduction of the CBP and research work was to lead to the greater use of some ingredients, such as enzymes. The greater potential for fungal alpha-

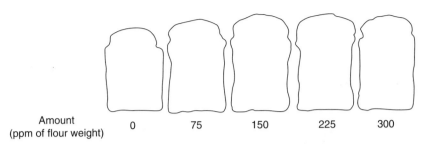

| Amount (ppm of flour weight) | 0 | 75 | 150 | 225 | 300 |

Fig. 2.4   Effect of ascorbic acid in the CBP.

amylase (Cauvain and Chamberlain, 1988) and other enzymes, including hemicellulases and lipases (Cauvain, 1985) was reported by the Chorleywood team in the 1980s.

In the UK the arrival of the CBP allowed millers and bakers to reduce the proportion of Canadian wheat used in flour grists. This action provided additional impetus to the desire to develop improved UK wheat varieties so that today it is not uncommon for CBP flours in the UK to be derived from 80–100% UK wheats. The change in the wheat composition used in UK milling grists was dramatic and by 1981 (Chamberlain, 1983) the reduced reliance on imported wheats was considered to have contributed, together with increased bread unit yields, to a reduction in bread costs of some £45 000 000.

The significance of the development of the CBP for British industry was recognised in 1966 with the granting of a Queen's Award to Industry for technical innovation to BBIRA. Despite the award the advent of the CBP went largely unremarked by the British public. Perhaps this was because the introduction of the CBP left UK bread quality relatively unchanged. The same could not be said for the quality of bread derived from American-style continuous mix process products. While the British public largely accepted CBP-type bread without comment, the same was not true for many bakers in the UK industry. Eventually the criticism from some bakers would spill out into the public domain and give rise to the myths that still exist in people's minds today about what the CBP is/was and what it is doing/has done for bread quality. Some of the erroneous issues raised are discussed in Chapter 3.

When launched, the CBP was intended for the small baker because the BBIRA team expected the plant baker to stay with continuous mixers. However, the plant bakers quickly recognised that the new batch system provided them with a sure way of reconciling raw materials into the plant with baked product coming from it. The introduction of the CBP eliminated the need for bulk fermentation periods with considerable savings in raw materials, time and space. It also initiated changes in ingredient and processing technologies which are still evolving today. The principles of the

CBP were to be adopted in many countries around the world (Gould, 1998). Even in those bakeries which did not adopt the CBP there has been a trend away from long periods of bulk fermentation to shorter processing times and the use of more functional ingredients to achieve more consistent bread quality (Cauvain, 1998a).

While the use of the CBP spread rapidly to many parts of the world, including South Africa, Australia, New Zealand and India, its potential introduction into the USA was less successful. The difficulties associated with using the CBP in the USA were linked with many factors, not least because of the higher work levels required to develop strong North American wheat flours which resulted in problems with higher than required dough temperatures at the end of mixing. In the 1980s the jacketed Tweedy mixer was patented following work at Chorleywood (Tweedy of Burnley Ltd, 1982) and this met the needs of temperature control for North American bakeries while delivering the higher energy levels needed for North American flours (Tomlinson, 1983) and achieving the fine cell structure and soft crumb required in the bread (Collins and Tomlinson, 1983). However, despite several attempts the general market in the USA was still not ready to adopt the CBP with a few exceptions.

Research into the 'new' breadmaking process continued when the Flour Milling and Baking Research Association (FMBRA) was formed from an amalgamation of the BBIRA and the Research Association of British Flour Millers. The breadmaking research team identified the critical role of oxygen and its affect on ascorbic acid functionality. They began to explain the role of partial vacuum in controlling the fineness of bread cell structure and with it the limiting effect on ascorbic acid functionality. To add to ascorbic acid functionality the team developed a modified form of the CBP in which the mixer headspace atmosphere could be oxygen enriched (Collins, 1986). The enrichment of the mixer headspace gases with oxygen allowed bread to be made with only ascorbic acid and without the application of partial vacuum with comparable quality to that seen with a mixture of ascorbic acid and potassium bromate and partial vacuum, the latter having become the UK industry 'norm'. This new development in the mid-1980s was to anticipate the potential loss of potassium bromate as a permitted oxidising agent. An 'oxygen-enriched' CBP plant was introduced but the concept was to fall foul of UK bureaucracy which considered that oxygen as such was not on the permitted list of ingredients for bread making and, thus, a process which used a mixture of 60% oxygen and 40% nitrogen was not permitted. Given this and other potential barriers to its implementation the approach was discontinued.

When potassium bromate was withdrawn from the permitted list of UK bread ingredients in 1990 the UK industry began to experience some bread quality problems. Delaying the application of the partial vacuum to the later stages of mixing and reformulation of bread improvers went some way towards solving the quality problems. However, it was clear that all was still

Fig. 2.5   T.H. (Bill) Collins at work in the Bread Bakery at FMBRA, Chorleywood, in the 1980s.

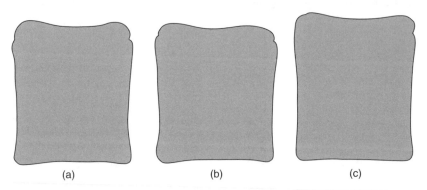

(a)   (b)   (c)

Fig. 2.6   Comparison of bread made on (a) a spiral mixer, (b) a spiral mixer followed by a Tweedy mixer at atmospheric pressure and (c) a spiral mixer followed by a Tweedy mixer with partial vacuum.

not right and the FMBRA bread team, still led by Bill Collins (Fig. 2.5), evolved the concept of the pressure–vacuum mixer (Collins, 1993). It was reasoned that what was required was the incorporation of large quantities of air to supply the oxygen for the ascorbic-acid-assisted oxidation and that could be achieved by mixing under positive pressure. However, the resultant crumb cell structure was coarse and open, more like a baguette. Clearly the application of partial vacuum was also needed to control bread cell structure.

The principles of a pressure–vacuum mixer were established by mixing dough for half its required energy using a spiral mixer and then manually transferring the dough to a Tweedy mixer for the remaining energy input and application of partial vacuum. This approach used the high gas occlusion properties of spiral mixers (Marsh, 1998) and controlled crumb cell structure partial vacuum on the Tweedy mixer. After filing a patent the FMBRA team set about persuading a mixer company, namely APV, to build a prototype machine which could be pressurised to 4 bar and move sequentially from positive to negative pressures during the mixing cycle. Examples of the different shapes and volumes of breads made by the various techniques at this time are illustrated in Fig. 2.6. The versatility of the machine was quickly apparent as products as diverse in cell structure as sandwich bread and baguette could be made with the same formulation and plant simply by varying the mixer headspace pressure.

The introduction of the CBP was undoubtedly a major event in the breadmaking world. It came about because of a mixture hard work and innovative thinking. The 1960s saw the arrival of many new breadmaking processes but the CBP alone has stood the test of time and because of this durability it has been a case of continued development from its first inception to the present day and will undoubtedly extend into the future.

3

Chorleywood Bread Process fundamentals

The main principles of the Chorleywood Bread Process

The CBP may be broadly described as 'a no-time dough-making process which uses mechanical development'. The basic principles involved in the production of bread and fermented goods by the CBP remain the same as those first published by the Chorleywood team in 1961 (Chamberlain *et al.*, 1961) although the practices have changed slightly with developments in ingredients and mixing equipment.

The essential features of the CBP (each of which is considered in more detail in the following sections) may be described as follows:

1 Mixing and dough development in a single operation lasting between 2 and 5 min to a defined energy input. Originally this was considered to be a fixed value equivalent to 0.4 hp min/lb, 5 W h/lb, 11 W h/kg or 42 kJ/kg dough in the mixer. While later work has shown that higher energy levels are required for some flours, optimum results (i.e. greatest product volume and finest cell structure) are obtained when the total required energy is delivered within the originally specified time. In the case of North American flours the upper limit of mixing time was extended to 7 min.
2 The addition of an oxidising agent, now restricted to ascorbic acid in many parts of the world.
3 The inclusion of a high-melting-point fat, emulsifier or fat–emulsifier combination.
4 The addition of extra water to adjust dough consistency to be comparable with that obtained with doughs produced by bulk fermentation. This extra water yields doughs with similar machinabilities which can

be processed on the same plant as bulk fermented doughs. While some aspects of plant design have changed since the CBP was introduced, the same principle related to dough consistency remains true.

5   The addition of extra yeast to maintain final proof times comparable with those seen with bulk fermentation doughs. The extra yeast is needed in CBP doughs because of the lower gas levels in the dough compared with bulk fermented doughs when they reach the start of proof.

6   The control of mixer headspace atmosphere to achieve given bread cell structures. When the CBP was first introduced, this was restricted to the application of partial vacuum for the whole or part of the mixing cycle but more recently this has been extended to include pressures greater than atmospheric and sequential changes during the mixing cycle.

The main difference between the CBP and bulk fermentation processes lies in the rapid development ('maturing') of the dough in the mixer rather than through a prolonged resting period. The advantages gained by changing from a bulk fermentation system to the CBP include the following:

1   A reduction in total processing time by at least the fermentation time of the dough in bulk.

2   Space savings from the elimination of the need to keep bowls of dough at different stages of bulk fermentation. These savings have been estimated as being a reduction in the mixing room area of 75% (Chamberlain *et al.*, 1961).

3   A reduction in the need for temperature-controlled areas for the bulk doughs with consequent energy savings. Although CBP doughs typically have a higher dough temperature than bulk fermented doughs, the shorter processing times and poor heat conductivity of dough mean that variations in ambient bakery temperature will have less impact on the dough temperature before reaching the prover.

4   Improved process control and reduced wastage in the event of plant breakdowns because there will be less dough at an intermediate processing stage, e.g. in bulk fermentation.

5   More consistent dough and final product quality through the elimination of the potential variability of dealing with doughs at different stages of fermentation.

6   More consistent dough and final product quality arising from the reduced variability in dough piece weights coming from the divider. This arises because CBP doughs have less gas in them at this stage than those made by bulk fermentation and the yeast is less active.

7   Financial savings from higher dough yield through the addition of extra water and retention of flour solids normally fermented away.

The disadvantages include the following:

1   The need to process the dough at a faster rate because of the higher dough temperatures used compared with those used with bulk fermentation.
2   A need for larger quantities of refrigerated water to control final dough temperature during mixing.
3   A second mixing is required for the incorporation of fruit into fruited breads and buns.
4   A possible reduction in breadcrumb (but not crust) flavour because of the shorter fermentation times involved in processing the dough to bread.

Definitely not the Chorleywood Bread Process

Like many new bakery-related inventions the CBP was not universally accepted when it first appeared in the baking industry. It had its detractors back in 1961 and it still has its detractors today. When Bill Collins retired from FMBRA, Chorleywood, in 1993 he said that one of the saddest things for him was that after 30 years he was still trying to teach people the basic principles of the CBP. Herein lies the Achilles' heel of the CBP, namely that the lack of understanding of the basic principles of breadmaking and the place of the CBP within the family of breadmaking processes has led to many misconceptions about the process and the products that may be made from it. As time passed, the early criticisms of the CBP passed into history and as such suffered the distortions which come from imperfect recollection and retelling. Now the misconceptions about the CBP have become the basis of myth and legend. A few examples follow:

1   A verbal comment made by a domestic science teacher after having had a guided tour of the bread test bakery at Chorleywood: 'Ah, yes, but I like my bread made with yeast!'
     No comment!
2   From http://www.thinkindustries.co.uk: 'But the Chorleywood Bread Process, invented in 1961, lets bakers skip those annoying few hours. The secret ingredient? A super high-speed, computer controlled mixer which sends the process into over-drive in about three minutes.'
     Extremely clever those computers in 1961!
3   From http://www.stylesource.co.uk: 'During the 30s, these ovens [coal-burning brick] started to be replaced with gas ovens and the Chorleywood Bread Process was introduced' and 'By the 50s and 60s the Chorleywood Bread Process was common, it's not really baking bread, more steaming it. . . .'
     Someone needs a history lesson.
4   From http://www.naturallaw.org.nz: 'Many processing agents are now produced using genetic engineering. For example bread is now

produced in UK and New Zealand using the Chorleywood Bread Process.'

The CBP really has a lot to answer for.

5   From 'Loafer in the lighthouse', *The Guardian*, 24 May 2003: 'Factory bread mostly uses the Chorleywood process with special high-speed mixers and flour "improvers" which produce an extremely light white loaf like a sponge which takes up lots of water.'

Does it come in any other colours? White or brown does not go with my bathroom décor.

6   From www.uktvfood/equipment/index: '*Bread machines* are a mini version of the Chorleywood industrial breadmaking process (sic), the method used to produce most bland mass-market bread available. . . .'

All that technology to make one loaf at a time!

7   Extract from an article published in the *British Baker*, entitled 'What have they done to our grain?': '*The Master Bakers Book of Bread-making*, published in the 1980s, prescribes up to 1.75% yeast in the dough made by the Chorleywood process. Could it be that by trebling (sic) yeast dosage and greatly reducing fermentation times, modern bakers have unwittingly laced their loaves with undesirable yeast residues? Some 40 years of Chorleywood bread amounts to a comprehensive experiment on the population.'

And the result is?

8   From *The Guardian*, Monday, 3 May 2004, the following extracts from *It's on the Label* (Lawrence, 2004):

(a) 'Researchers at the British Baking Industries Research Association in Chorleywood developed a way of making bread that had first been used in the US. . . .'

Certainly, using mechanical development was recorded in the USA in the 1920s but was never used until after the launch of the CBP.

(b) '. . . instead of allowing two to three hours of fermentation they found that air and water could be incorporated into dough if it was mixed at high speed in mechanical mixers.'

(c) 'Air and water are the secrets of that springy loaf.'

So what did bakers use before the invention of the CBP?

(d) 'The actual breadmaking is over in the blink (of an eye). This is the wonder of the CBP.'

If only!

(e) '. . . but for CBP flour you do an enormous amount of grinding under huge pressure, to break open all the starch molecules (sic). The flour ends up very fractured and grey.'

If every starch granule, sorry molecule, is 'broken,' just imagine the water absorption of the flour!

9   From David (1977): 'By the mid 1960s the notorious Chorleywood Bread Process had been evolved. Unknown to the general public

the instant (sic) factory loaf had superseded the plant loaf of the 1950s.'

   If only, did I hear bakers cry?

10   From Keating, *Sunday Times Magazine* (2005): 'If you use high-speed mixers, the process is so fast that the gluten and yeast aren't broken down properly. It's a bit like eating raw dough, or bread straight from the oven, because everything is still working. . . .'

Even reputable works which refer to the CBP are prone to misunderstandings and errors in reporting details, as in the following two examples:

1   Connolly and Pittman (1997) reported 'The level of water must be increased by approximately 2% of the flour weight. This is because the flour is able to absorb more water during mixing.'

   While it is true that water levels in CBP doughs are 2% or more higher than would have been in bulk fermentation systems, the increase arises from two sources: one is the need to adjust dough consistency by extra water addition to compensate for the lack of softening which occurs during bulk fermentation and the other is to compensate for the firmer dough which results from the application of partial vacuum during mixing.

   The same flours may be used in the CBP as may be used in any other breadmaking process. The misunderstanding arises because many flours used in the production of plant-type breads have higher water absorptions because of their higher starch damage levels and it is the design of the total plant which permits the dough to be processed with higher added water levels and not just the use of CBP mixing techniques.

2   Davidson (1995) perpetuated the myth that fermentation is not part of the CBP by confusing lack of 'bulk' fermentation with no fermentation. As with many other authors the confusion arises because of the terminology used by bakers which gives the impression that 'proof' is not associated with fermentation. Given that proof is commonly carried out at much higher temperatures than bulk fermentation, then the products of yeast fermentation are considerable. What is less favoured during proof is the fermentation which comes from other micro-organisms present in the dough system, e.g. lactobacillus, when the dough is fermented at the lower temperatures used with bulk fermentation processes. Lactobacilli are naturally present in the flour and the fermentation processes concerned generates acid notes in the final product. This effect becomes particularly evident with the retarding of fermented dough.

Other no-time dough development processes

There have been many other no-time and mechanical dough development processes. The differences between those which remain in common use today and the CBP may be summarised as follows:

Amflow dough production

Amflow dough production uses a pre-fermentation system (typically 2–2.5 h) and continuous mixing. This process may incorporate a degassing unit before dividing to aid divider weight control. There is no direct control over crumb cell structure other than through degassing. It is commonly used in the production of hamburger buns (Bent, 1998).

The main difference from the CBP lies with dough production by continuous rather than batch mixer and the use of a degasser. The latter is required because the pre-fermentation stage yields a dough with considerably more gas than typically seen with CBP doughs.

Mechanical dough development in Australia

Mechanical dough development in Australia was first seen as the Brimec process developed by the then Bread Reseach Institute (BRI) of Australia and launched in 1962. This process featured dough development in the mixer with some crumb cell structure control by varying the position of a ram which restricted the free space in the mixing chamber and exerted some pressure on the dough. The dough could be shaped and placed directly in a pan.

A no-time dough process using extra mixing on low-speed mixers was launched in 1964 (Collins *et al.*, 1968). In 1965 an extended proof version of mechanical dough development was launched by the BRI with low yeast level and long proof times (typically 16–18 h).

Today mechanical dough development in Australian bakeries is very similar to the CBP, other than the tendency to use higher work levels in the dough because of the generally stronger wheat varieties which are available in Australia and New Zealand.

Spiral mixing

The range of processes covered by spiral mixing are commonly seen in smaller bakeries and those in which the finest breadcrumb cell structure is not sought. The mixing action is based on one or two spiral-shaped beaters, sometimes with a fixed bar also in the bowl (Marsh, 1998). Energy input in a given time is normally lower than that achieved in a CBP-type mixer. Mixing conditions are controlled by time and there is no direct control of crumb cell structure. The dough is processed within a few minutes of the completion of mixing.

In the UK and Ireland, spiral mixing may be seen as a descendant of the activated dough development (ADD) group of breadmaking processes. Such processes used low-speed mixers and a mixture of potassium bromate, ascorbic acid and L-cysteine (Collins *et al.*, 1968). When potassium bromate was removed from the permitted list of oxidising agents in bread and prices of L-cysteine began to rise steeply, ADD processes went out of use.

The main differences between spiral mixing processes and the CBP lie in the longer mixing time, lower rates of energy input and the lack of crumb cell structure control. This means that the range of bread types which can be made using this process are restricted to more open cell structure products, such as baguettes.

Sponge and dough in the USA

The sponge and dough process in the USA is based on pre-fermentation of part of the dough recipe ingredients, typically for up to 4 h. The sponge is added to the rest of the ingredients and mixed to a dough in a horizontal bar mixer. Control of dough development is through the mixing time, and the rates of energy transfer are lower than those achieved in the CBP. The greater mixer surface area and longer contact times between dough and mixing bowl permit better use of a cooling jacket to control the final dough temperature. There is no direct control of crumb cell structure and a degasser may be used to achieve improved divider weight control.

Mixing times are longer than with CBP because of the lower rate of work input. Gas levels in the dough when it leaves the mixer are much higher. Degassing compensates in part for the lack of crumb cell structure control and so finer crumb cell structures are possible.

4

Ingredient qualities and quantities in the Chorleywood Bread Process

Flour

No special flours are required for the manufacture of fermented products by the CBP. However, as discussed above, the introduction of the CBP permitted a reduction in flour strength below that used in bulk fermentation processes of the time for the production of breads of limited volume (e.g. sandwich loaves). This has led to the misconception that weak flours are an essential requirement for the manufacture of bread by the CBP.

It is now fully accepted that the 'optimum' energy level used for dough mixing in the CBP varies with wheat variety and flour strength. In general, the optimum energy required for dough development in the CBP increases with increasing flour protein content. An example of the link between flour strength and energy expenditure during mixing is illustrated for a 'strong' wheat variety in Fig. 4.1. As the level of energy expenditure during mixing increases, so does the volume of the resultant bread. Maximum bread volume for this example is reached at around 17 W h/kg dough in the mixer, somewhat higher than the conventional 11 W h/kg. Wheat varieties which yield flours with lower protein contents may achieve optimum bread volume at lower energy expenditures. In some parts of the world, this may be difficult to achieve if lower-protein wheats are not available. Each wheat variety will yield flour which exhibits a slightly different pattern from that shown in Fig. 4.1. In practice, the UK flour miller will blend wheats to give a flour requiring an energy expenditure of around 11 W h/kg. This approach is illustrated in Fig. 4.2 where a 'strong wheat', namely Fresco, has been blended with a 'weak' wheat, namely Riband, in the proportion of 3:1 for the production of flour and the 'optimum' work input occurs at around 11 W h/kg.

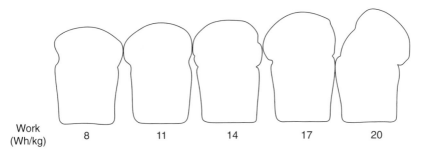

Work
(Wh/kg)   8   11   14   17   20

Fig. 4.1   Effect of increasing work input with a flour from a 'strong' wheat.

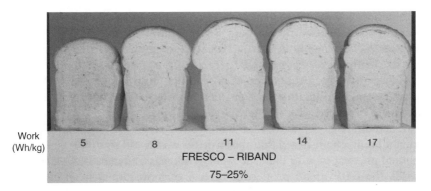

Work
(Wh/kg)   5   8   11   14   17

FRESCO – RIBAND

75–25%

Fig. 4.2   Effect of increasing work input with a flour produced from blended strong
and weak wheats.

Protein content alone is not the sole factor which determines the
optimum energy input level in the CBP. The effects of protein quality are
often less clear in the CBP than with some other breadmaking processes,
in part because of the relatively high level of oxidants used in the recipe
which influences dough rheology and in part because dough-processing
times from mixer to prover are much shorter. It is therefore difficult to
relate common dough rheology tests to the CBP bread quality. This is not
surprising given that the common protein quality tests which are used are
not related to dough-mixing conditions in the CBP. One protein quality test
which does appear to have some relationship with CBP bread quality is the
'gel protein test' (Alava *et al.*, 2001) which has been used to identify wheat
and/or flour strength. However, the range of 'acceptable' gel protein values
is probably too broad for the test to be used without interpretation by
appropriate 'experts' and the test has not found universal acceptance.

Measurements of changes in dough rheology have been made using both
mixer torque and through the application of near-infrared (NIR) spectroscopy to dough mixing using a CBP-type laboratory mixer (Millar, 2003).
Both methods of evaluating dough properties have confirmed that a rela-

Fig. 4.3   Effect of increasing flour protein content in the CBP.

tionship exists between optimum energy input during mixing and the strength of the wheat flour used. NIR data derived from curves in the region 1125–1180 nm have been used to relate flour properties to the elastic modulus of the gel protein fraction of flours (Alava *et al.*, 2001) and differences in gluten qualities which clearly relate to flour performance have also been observed with this technique.

When different wheats are mixed to provide a grist for milling CBP-type flour, it has been observed that the stronger wheat will exert a greater effect on final product quality than the weaker wheat (Cauvain, 1998a). In other words, the CBP bread properties of two or more flours from individual wheat varieties cannot be predicted by the weighted average of their individual properties. The effect of flour protein content with a fixed energy expenditure during mixing is illustrated in Fig. 4.3 and shows that, as flour protein increases, so does bread volume (as would be expected in most other bread-making processes). It should be noted that the loss of bread volume observed with the 11.6% protein flour could be compensated for by adjusting the ingredients in the improver formulation. The protein content of wheat flour can be adjusted through the addition of dry vital wheat gluten (Chamberlain, 1984). There appear to be no problems associated with the use of gluten-supplemented wheat flours in the CBP and full use is made of the added gluten protein (Collins and Evans, 1984; Collins *et al.*, 1985; Collins and Fearn, 1986).

The presence of bran in white flour reduces the ability of the dough to retain gas in the CBP (Cauvain, 1987) and indeed in any other bread-making process. In general, the higher the level of bran, the poorer is the gas retention in the dough and the smaller is the bread volume. To compensate for the higher bran content the protein content of the flour can be raised using higher-protein wheats in the milling grist or through gluten supplementation (Collins and Young, 1986). The latter approach is common with UK wholemeal flours and the efficacy of such fortification has already been commented on above.

The Hagberg Falling Number has particular relevance in CBP-type flours. Low Hagberg Falling Numbers indicate a high level of cereal *alpha-*

amylase activity. While cereal *alpha*-amylase does improve gas retention in the dough, it also has potential negative effects. Dextrins are produced by the action of cereal *alpha*-amylase on damaged starch during proof and baking and they are sticky in nature. If their level in the baked bread is high enough, they can cause a build-up of residues on the slicer blades in the plant which leads, in turn, to reduced slicing efficiency and crushing of loaves. The effects of cereal *alpha*-amylase in the CBP have been extensively studied (Chamberlain *et al.*, 1977) and a minimum Hagberg Falling Number is an important part of the flour specification. The minimum Falling Number which can be tolerated in the bakery plant varies according to many factors but, in general, lower values are tolerated than when the CBP was first introduced. This change probably has much to do with changes in recipe, improvements in plant operation and slicing efficiency.

During the flour-milling process a proportion of the starch granules which are present in the endosperm become mechanically damaged. Such starch granules make a significant contribution to the water absorption capacity of the dough. Damaged starch granules absorb about twice their weight of water while intact granules only absorb about 0.4 times their own weight. Thus, the higher the level of damaged starch in the flour, the higher will be its water absorption capacity, although there is some evidence that bread volume may be reduced with increasing levels of added starch damage (Fig. 4.4). The damaged starch levels in CBP-type flours have increased significantly since its introduction in 1961.

There is an important link between the level of damaged starch and the presence of *alpha*-amylase in the flour, whether naturally occurring or added in the flour mill. The presence of both *alpha*- and *beta*-amylase in the flour leads to the breakdown of the starch, first to dextrins (as noted above) and ultimately to maltose (a fermentable sugar; see the next section

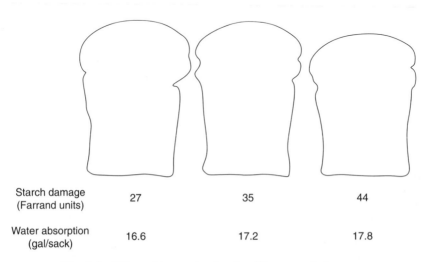

| Starch damage (Farrand units) | 27 | 35 | 44 |
| Water absorption (gal/sack) | 16.6 | 17.2 | 17.8 |

Fig. 4.4   Effect of increasing levels of flour starch damage.

on yeast). However, for the *alpha*-amylase to work the starch granules must be damaged, such as would result from the milling of wheat to flour.

Flour properties other than damaged starch affect the water absorption capacity of wheat flour. These are mainly as follows:

1 *The flour moisture content.* The water absorption capacity increases as the moisture content falls.
2 *The protein content.* The water absorption capacity increases as the protein content increases. Protein absorbs about 1.3 times its own weight of water.
3 *The pentosan content.* The water absorption capacity increases as the pentosan level increases. Pentosans absorb about seven times their own weight in water but are only present at relatively low levels, typically about 2% of the flour weight.
4 *The bran content.* The water absorption capacity increases as the level of bran increases.

There is some evidence that flour particles in the range 50–70 μm increase water absorption (together with bread volume and crumb softness) but generally particle size is not considered to be an important flour characteristic for the CBP. However, it should be noted that the hydration of flour particles is not an instantaneous event and so the rate at which water penetrates the larger endosperm particles may affect gluten development in the short mixing times which characterise the CBP.

One flour property not commonly considered in bread making but which has important implications in the CBP is known as 'heat of hydration'. Essentially when flour and water are mixed together the hydration of the flour results in a release of energy which contributes to dough temperature rise, independent of energy transfer. The property is directly related to the flour moisture content (Wheelock and Lancaster, 1970). In the CBP the additional temperature rise generated by the hydration of flours during mixing may cause further problems with final dough temperature control.

The range of typical white-flour properties which might be used in the CBP are summarised in Table 4.1. The commonly applied terminology in the UK would be 'CBP grade' for protein contents around 11% and 'bakers'

Table 4.1   White-flour properties for CBP bread production

Flour property	Typical range of values
Moisture (130 °C for 1.5 h) (%)	13.5–14.5
Protein (as is) (%)	10.5–12.5
Grade colour figure	0–2.5
Falling number (s)	250–350
Damaged starch (Farrand units)	25–45
Water absorption (Farinograph) (%)	58–62

grade' for protein contents around 12%. Precise specifications will vary according to the needs of the baker. In principle, any flour may be used for the manufacture of bread using the CBP. Wholemeal and other speciality flour tend to have higher protein contents to boost dough gas retention.

Yeast

Wheat flour for breadmaking contains about 1% naturally occurring fermentable carbohydrates which are available for yeast fermentation. These consist of sucrose, glucose, fructose and a series of polysaccharides. When flour and water are mixed, maltose is produced by the action of the amylases which are naturally present, or which have been added in the flour or the improver. When presented with the mixture of sugars in the flour, the yeast preferentially ferments any glucose and fructose present and does not begin fermentation of the maltose until the monosaccharide sugars are exhausted. One consequence is that the level of maltose in the dough rises during the initial stages of any fermentation through the action of the amylases.

It is not common to add sugars to CBP bread recipe in many parts of the world and so yeast strains which adapt rapidly to maltose fermentation are best suited to the process. The levels of yeast used in the CBP are typically higher than seen in bulk fermented doughs. The higher yeast level is used to maintain final proof times to be comparable with those seen with bulk fermentation processes. This occurs in part because CBP dough contains considerably less gas when it arrives at the divider than is seen with bulk fermentation systems, typically around 10% in the CBP and around 70% in bulk fermented doughs. Even though some degassing will occur when bulk fermented doughs are divided, the gas levels in such doughs remain much higher than seen in CBP doughs. Since CBP dough has a lower gas volume, then, for a given dough and prover temperature, it takes longer for it to achieve a given proof volume before baking and therefore to compensate for this yeast levels are raised.

In doughs which contain no added sugar, yeast strains which adapt slowly to maltose fermentation will show a marked drop in the rate of gas production towards the end of final proof. When first introduced, problems were experienced with lack of oven spring in CBP doughs. In the early 1960s the yeast in common use was relatively slow acting. This was not a problem in bulk fermentation because processing times were long enough for the production of maltose from the action of amylases on the damaged starch in the flour. CBP doughs typically reached the oven about 60–70 min after the start of mixing and this roughly coincided with a slowing down of yeast activity and in some cases a fall in gas production rate (Williams and Pullen, 1998). The solution was to change the strain of yeast used so that there was no significant loss of gas production at the critical moment of the doughs entering the oven.

An important consideration in the CBP is the significant oxygen-scavenging potential of yeast. This means that, even in the short mixing times which are common in the CBP, little or no oxygen remains in the dough by the end of mixing. Cauvain (unpublished observations) has shown that the volume of air in a sealed mixer reduces by approximately 20% (the approximate concentration of oxygen in air) by the end of mixing with the CBP. In the absence of oxygen the conversion from ascorbic acid to dehydroascorbic acid cannot take place and this restricts the potential for ascorbic acid oxidation. The importance of this situation is discussed in the section on oxidants below.

All forms of yeast may be used with the CBP. In larger plant bakeries it has become common to use liquid or cream yeast which can be pumped directly and metered automatically to the mixing bowl. In smaller bakeries it is more common to use compressed yeast. Dried and pelleted yeasts may also be used provided that the manufacturer's recommendations are followed.

Salt

In the CBP the basic functions of salt in bread doughs are to contribute flavour, to control fermentation (Williams and Pullen, 1998) and to influence dough rheology. No special types of salt are required with the CBP, although the salt particle size should be small enough to ensure complete dissolution within a couple of minutes of starting mixing.

Water

The 'optimum' level of water in bread making is in effect the maximum quantity that can be added which will yield a dough that can be processed on the plant and give bread of acceptable quality. The optimum water level to be used in the dough depends heavily on the flour properties, as discussed previously. In general, the water levels added to CBP doughs are usually higher than with other breadmaking processes. The addition of extra water is required to adjust dough consistency to provide a softer, more machinable dough similar to that obtained with bulk fermentation doughs. This requirement occurs in part because CBP dough contains less gas than bulk fermented dough at the point of dividing and in part because there is limited time for enzymic softening compared with bulk fermentation.

The requirement for extra water is particularly true when the dough is mixed under partial vacuum in the CBP (Cauvain, 1998a). The lower the pressure during mixing, the 'drier' the dough feels and the more water needs to be added to achieve the same dough consistency for processing. The reverse is also true in that mixing doughs under positive pressure increases the volume of occluded air and yields a softer dough which may require some compensatory reduction in added water levels.

Oxidants

One of the basic principles of the CBP is the addition of an oxidising agent to improve the gas retention abilities of the dough. The functions of the oxidant are complex and depend on the particular oxidising agent that is being used. However, the main effect of any oxidant is to increase dough gas retention, generating greater oven spring and increasing loaf volume.

The oxidants which have been or may be used in the CBP are listed in Table 4.2 together with typical ranges for levels of use and comments on their current legal status in some countries.

Potassium bromate was a commonly used oxidant in the early days of the application of the CBP. Its value lay in the delay of most of its activity until the later stages of proof and the early stages of baking. In the CBP it was most commonly used in combination with ascorbic acid because of the synergistic effect between the two oxidants when mixed in air at atmospheric pressure (Chamberlain and Collins, 1981). The degree of the synergistic effect varied with mixer type.

The use of ascorbic acid in bread making is not as straightforward as that of other oxidants such as potassium bromate. Chemically, ascorbic acid is a reducing agent and can only function as an oxidising agent in dough after it has been itself oxidised to dehydroascorbic acid (Melville and Shattock, 1938). The presence of ascorbic acid oxidase (an enzyme which occurs naturally in flours) and oxygen is required to achieve this conversion. The

Table 4.2 Oxidant types and their levels of use in the CBP

Oxidant	Typical level of addition (ppm flour weight)	Legal status
Ascorbic acid	75–300	Permitted in most countries
Potassium bromate	15–50	Limited use in the USA Not permitted in the EU, Australia, New Zealand, Canada
Azodicarbonamide	5–25	Permitted in the USA Not permitted in the EU, Australia, New Zealand, Canada
Potassium iodate	10–30	Permitted in the USA Not permitted in the EU, Australia, New Zealand, Canada

oxygen mainly comes from air trapped in the flour and from air bubbles incorporated into the dough during mixing. As will be discussed in Chapter 5, the quantity of gas incorporated into CBP doughs during mixing depends in part on the mixer headspace pressure.

In the CBP, the bread volume initially increases as the level of added ascorbic acid increases to reach a maximum value beyond which further additions of ascorbic acid have no effect. The point at which the maximum is reached depends on the availability of oxygen in the mixer. For example, the application of partial vacuum during mixing decreases the volume of air occluded and gives lower bread volumes; increasing the mixer headspace pressure increases the occluded air volume and bread volume (for a more detailed discussion of the effects of mixer headspace pressure in the CBP, see Chapter 5). When ascorbic acid is used alone, improvements in bread volume may also be obtained in the CBP by increasing the concentration of oxygen in the mixer headspace (Collins, 1986) (Fig. 4.5). In addition to affecting the bread volume, the use of ascorbic acid has an impact on dough rheology in the CBP. Cauvain *et al.* (1992) used large-scale deformation testing and showed that dough resistance to deformation when using ascorbic acid was greater than the effect of potassium bromate. The different effects on dough rheology from the mixer probably reflect the different rates of reaction of the two oxidants.

The availability of oxygen for the conversion of ascorbic acid to dehydroascorbic acid is an important consideration in choosing the oxidant to use in the CBP. As noted above, yeast is a scavenger of oxygen in dough and therefore tends to restrict the action of ascorbic acid. However, it should be noted that the oxygen-depleted situation which arises in the dough as it enters the prover or very soon after entry might provide opportunities for its action as a reducing agent, although such a scenario remains the subject of conjecture.

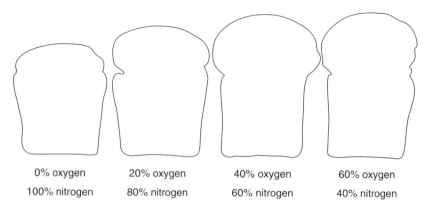

| 0% oxygen | 20% oxygen | 40% oxygen | 60% oxygen |
| 100% nitrogen | 80% nitrogen | 60% nitrogen | 40% nitrogen |

Fig. 4.5 Effect of oxygen concentration on bread volume.

Fig. 4.6   The effect of 100 ppm of ascorbic acid and 100 ppm of dehydroascorbic acid.

Since the known oxidising agent is dehydroascorbic acid, then the possibility exists for direct addition of the chemical itself rather than relying on the conversion in the mixer. Collins *et al.* (1991) obtained three samples of dehydroascorbic acid and evaluated their performance against that of a standard addition of ascorbic acid. Of the three samples, one was a commercially available sample (Aldrich Chemical Co.) and the other two were prepared in the laboratory. One of the two laboratory samples was fat coated. The Tweedy mixer headspace conditions used for the trials were as follows: atmospheric pressure, partial vacuum of 0.5 bar and a mixture of 60% oxygen and 40% nitrogen. The mixer headspace conditions were maintained throughout the whole of the mixing period. All three dehydroascorbic acid samples gave similar performances in terms of bread volume and crumb cell structure and the results were similar to those with ascorbic acid. The exception was the use of dehydroascorbic acid with partial vacuum when improvements in bread volume, crumb structure (finer), softness and crumb colour compared with ascorbic acid were observed. The effects of 100 and 200 ppm ascorbic acid and dehydroascorbic acid are shown in Fig. 4.6 and Fig. 4.7 respectively.

Arguments have also been advanced for the generation of hydrogen peroxide in the proving dough through enzyme catalysis. It should be remembered that in both of these speculative scenarios the release of oxygen is into a dough system in which the components have limited mobility. This contrasts with the situation in the mixing of dough where 'new' air is constantly being incorporated. Thus, any impact of either the ascorbic acid reduction reaction or the generation of hydrogen peroxide will be small by comparison with the level of ascorbic-acid-assisted oxidation in the mixer.

In the past this dependence of ascorbic acid on the available oxygen has led to the view that it is not possible to 'over-treat' with ascorbic acid in the CBP but, by increasing the availability of oxygen, over-treatment can be

No ascorbic acid   200 ppm ascorbic acid   200 ppm dehydroascorbic acid

Fig. 4.7   The effect of 200 ppm of ascorbic acid and 200 ppm of dehydroascorbic acid.

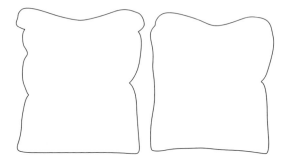

Fig. 4.8   Collapse in lidded bread from excessive gas retention.

demonstrated with some bread types. For example, the lidded loaves in Fig. 4.8 show that by increasing mixer headspace pressure with ascorbic acid in the dough there is a tendency for sidewall and top-crust collapse to occur. This phenomenon is most often associated with a 'weakness' in the dough but is more likely to come from over-expansion of the centre crumb because of good (excessive) gas retention (Collins and Sang, 1993). In the case of the loaves in Fig. 4.8 this occurred because of improved ascorbic-acid-assisted oxidation during mixing.

In some countries (e.g. the USA), potassium iodate, potassium bromate, azodicarbonamide and other oxidants remain available for addition by the miller and baker and these may be used with the CBP to replace ascorbic acid. Such oxidants tend not to be as oxygen sensitive as ascorbic acid and so are less affected, if at all, by mixer atmosphere or headspace pressure. For example, Chamberlain and Collins (1984) showed that the effect of potassium bromate in the CBP was not oxygen dependent in commercial bakery trials using different mixers and different atmospheric condition in the mixer headspaces. However, Chamberlain and Collins (1979) did show that, in the absence of oxygen in the mixer, azodicarbonamide was less effective as an oxidant in the CBP.

Each oxidant will have a different rate of reaction in the dough. For example, potassium bromate is known to react more slowly with most of the oxidation effect taking place towards the end of proof and in the early stages of baking, while azodicarbonamide, potassium iodate and ascorbic acid are much faster acting with much of the oxidation occurring in the mixer, even with the short mixing times encountered in the CBP.

Reducing agents such as L-cysteine may be added to 'weaken' the dough structure or to modify its rheology in order to improve moulding and shape forming, such as with rolls and baps, without structural damage. L-cysteine and other reducing agents are sometimes recommended when using very strong (high protein or 'tough' gluten, or both) in the CBP when it is not possible to deliver the high levels of energy that would normally be required for full development of the dough. In some cases, the addition of reducing agents is used in an attempt to limit final dough temperature by reducing the energy input during mixing. In all cases the benefits of adding reducing agents to adjust bread quality are equivocal.

Fats

Compound bakery fats are a mixture of oil and solid fat at a given temperature. The proportion of solid fat present at a given temperature may be expressed as the solid fat index (SFI) or solid fat content (SFC). In the past it was more common to refer to the melting point of the fat, which is the temperature at which the fat became 100% oil (Stauffer, 1999).

As early as 1942 the bread improving of fat with 'no-time' (i.e. no bulk fermentation or floortime) was recognised (Baker and Mize, 1942). The CBP is a no-time dough-making process and during its development it was recognised that the addition of fat improved the gas retention of the dough and thereby increased bread volume and softness (Chamberlain *et al.*, 1965).

Increasing the fat level in the recipe will increase bread volume up to a point and thereafter there will no further significant increase in volume for further additions. This level will vary according to the type of flour being used and so it is common to recommend a 'blanket' level in order to ensure that sufficient solid fat is always present. With white flours this is usually 0.7% flour weight or higher. Wholemeal (wholewheat) flours commonly require higher levels of fat addition, often twice to three times that of white flour to achieve maximum bread volume (Williams and Pullen, 1998).

It is important that a proportion of the fat should remain solid in bread dough at the end of final proof (Chamberlain *et al.*, 1965) typically at around 40–45 °C. The SFC profiles of some typical fats are illustrated in Fig. 4.9. The butter would not be suitable for use in CBP to improve gas retention although it may be added to modify flavour. The puff pastry fat would be

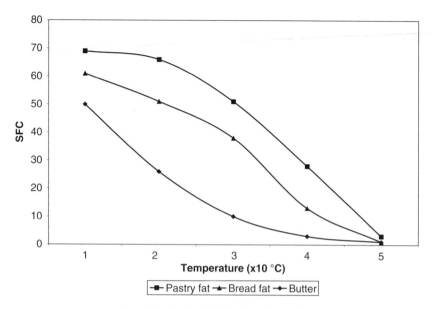

Fig. 4.9   Typical SFCs for bakery fats.

suitable for addition to improve gas retention but, because it has a very high proportion of solid fat at low temperatures (20–30 °C), problems with uneven dispersion may be encountered. Typically, composite bread fats will have a liquid oil component to aid dispersion and a solid fat component to aid gas retention. The fat crystal size also plays a part in the volume-improving effect, with small fat crystals having a greater improving effect than large crystals.

The mechanism by which solid fat improves dough gas retention with the CBP has been studied by a number of workers (Brooker, 1996). In the dough the solid fat crystals are mostly located at the interface of the gas bubbles with an aqueous phase in the matrix where they augment the bubble-stabilising properties of the gluten network and other functional ingredients, such as emulsifiers.

Emulsifiers

Emulsifiers may be added to bread to improve its quality, each one acting slightly differently and having its own special effects in the dough and the final product (Kamel and Ponte, 1993; Williams and Pullen, 1998). The four commonly used emulsifiers are the following:

1   Diacetyltartaric acid (DATA) esters of monodiglycerides (DATEM).
2   Sodium stearoyl lactylate (SSL).

Table 4.3 Typical levels of emulsifier addition

Emulsifier	Level of addition (% flour weight)
DATA esters (DATEM)	Up to 0.3%
SSL	Up to 0.5%
GMS (DGMS)	Up to 1%
Lecithins	Up to 0.5%

3   Distilled monoglyceride (DGMS) and glycerol monostearate (GMS).
4   Lecithins.

DATA esters and SSL may be used in the CBP to improve dough gas retention and dough stability and to yield a finer cell structure in the product, as
shown by Collins *et al.* (1985) for wholemeal bread and by Evans (1984)
for soft rolls. DGMSs are most commonly used to reduce crumb staling
(Pateras, 1998) and to improve crumb softness. Lecithins may be used to
improve crust crispness with crusty products, such as baguettes. Typical
levels of emulsifier addition in the CBP are given in Table 4.3.

Enzyme-active materials

Enzyme-active materials have become important to many sectors of the
baking industry following the limitations placed on the use of many oxidants. The mixing and processing times for the CBP are shorter than for
many other breadmaking processes and so the effects of enzymes on dough
rheology before the pieces reach the prover tend to be less obvious
(depending on the specific activity of the enzyme preparation, the level of
addition and the dough temperature). Those most commonly added
enzymes in CBP are the fungal (Collins and Evans, 1982; Cauvain *et al.*,
1985; Collins *et al.*, 1985; Cauvain and Chamberlain, 1988), cereal (Dodds
and Collins, 1971; Collins and Evans, 1982) and maltogenic (Williams and
Pullen, 1998) amylases. Other enzymes find their use in CBP breadmaking;
they include lipsases (Cauvain, 1985). Proteolytic enzymes may be used in
the USA (Kulp, 1993) or where strong high-protein flours are available.

Sugar

Sugars in crystalline or liquid form may be added to CBP doughs to modify
flavour and product colour. It is not necessary to add sugars to support yeast
fermentation in CBP because dough-processing times are too short. High
levels of sugar inhibit yeast activity but can be accommodated in practice

by raising the added yeast level. Even though the quantities of sugar which may be added may rise to as high as 30% flour weight (e.g. in southern India) the inclusion of sugar in the CBP usually presents no problems during dough production. The relatively large quantity of water in the dough and the relatively high dough temperatures achieved usually ensure full dissolution. Nevertheless it is usually advisable to avoid adding coarse-grained sugars, such as the granulated form.

Soya flour

Full-fat enzyme-active soya flour has been used as a functional dough ingredient in the UK since the 1930s. It has two principal beneficial functions in CBP, both arising from its lypoxygenase enzyme system. They are to provide a whiter breadcrumb by bleaching the flour and to assist in dough oxidation, which improves bread volume (Collins and Smith, 1976). Soya flour is commonly used as a component in improvers and concentrates.

Mould inhibitors and preservatives

In common with other breadmaking processes, mould inhibitors and preservatives may be added to CBP doughs in order to delay spoilage (Pateras, 1998; Williams and Pullen, 1998). In addition to their positive impact on bread spoilage, these materials can have negative technological impacts. The most common negative impact is on yeast activity and so it may be necessary to compensate for such an effect by increasing recipe yeast levels or extending final proof time, or both. Direct contact between mould inhibitors and yeast is best avoided. Encapsulation of mould inhibitors offers some potential in this latter respect since control of the release of the inhibitor can be delayed until after the main activity of the yeast has been achieved (e.g. above 45 °C).

Some typical inhibitors and their levels of addition are given in Table 4.4.

Table 4.4   Mould inhibitors and preservatives

Ingredient	Level of addition (ppm of finished product)
Acetic acid (vinegar, antirope)	*QS**
Sodium diacetate	2000
Propionic acid (antimould)	2000
Potassium, sodium or calcium propionate (antimould)	3000

* *Quantum Satis*, i.e. without limit, although the technological and sensory effects will limit the levels of addition.

Other ingredients

Many other ingredients can be used in the production of bread products by the CBP. These include the following:

1 *Milk powders.* These should be fully heat treated to avoid the potential loss of dough gas retention (Cauvain and Young, 2001). The water level added to the dough will need to be increased to compensate for the presence of the milk products.
2 *Wheat bran and other fibres.* These are commonly used in the production of brown, bran-enriched and 'white' fibre breads. The addition of extra water in the recipe is required to yield a dough of 'standard' consistency. On occasions this may be difficult to judge since bran and fibre materials may be slow to absorb water and so may not become fully hydrated until later in the dough-processing stages. In these circumstances the dough may appear soft or 'sticky' on leaving the mixer but will become firmer during moulding and resting.
3 *Fruits and nuts.* These should be added late in the mixing process in order to avoid breakdown of the particulate material. Sometimes this may require the transfer of dough to a separate mixer. This may be inconvenient but may be necessary to avoid smearing or 'bleeding' of fruit in the dough. If fruits or nuts are used in the formulation, they should not be included in the calculation of the required mixing energy level.

Improvers and concentrates

Improvers and concentrates are any ingredient added to 'improve' the breadmaking potential of a flour, i.e. to improve carbon dioxide gas production or to modify the gas-holding potential and the rheology of the gluten network formed in the dough. More than one improving or functional ingredient may be added at a time. Improver formulations are usually matched to a given flour, product, process or final product characteristic. Thus, there are many different improver formulations which may be used in the production of bread. The choice of ingredients which may be used is commonly limited by legislation in a given country.

Improvers of one form or another are used to deliver the required functional ingredients for the manufacture of fermented products using the CBP. The improvers may be used in the powder, paste or liquid form. Concentrates may include other recipe ingredients such as milk solids, fat and sugar and may also be used with the CBP. The improvers may be added by hand at the mixer or automatically. The options are discussed further in Chapters 5 and 9.

5

Dough mixing and the Chorleywood Bread Process

The functions of mixing

Mixing is the homogenisation of the ingredients, whereas kneading is the development of the dough (gluten) structure by 'work' done after the initial mixing. In the context of the CBP, both processes take place within a single mixing machine and so can be considered as one operation.

The subprocesses taking place during mixing were summarised by Cauvain (2001) as follows:

1 The uniform dispersion of the recipe ingredients.
2 Dissolution and hydration of those ingredients, and in particular the flour proteins and the damaged starch.
3 The development of a gluten (hydrated flour protein) structure in the dough arising from the input of mechanical energy by the mixing action.
4 The incorporation of air bubbles within the dough to provide the gas bubble nuclei for the carbon dioxide which will be generated by yeast fermentation and oxygen for oxidation and yeast activity.
5 The formation of a dough with particular rheological properties for subsequent processing.

Types of mixer

The essential features of the CBP and the evolution of the high-speed mixer that was required have been described in earlier sections. The most common CBP-compatible mixers consist of a powerful, vertically mounted motor drive, directly coupled through a belt system to a vertically mounted mixing

Fig. 5.1   CBP-type mixer – Tweedy 35, similar to that originally used in the CBP.

blade in a fixed cylindrical bowl (Fig. 5.1). During mixing, the dough is thrown outwards by the action of the impeller or mixing blade to hit the walls of the mixing chamber. After making contact with the sides of the mixing bowl, the dough mass falls to the bottom under the influence of gravity to be lifted again by the impeller. In the vertical CBP-type mixers there are projections from the inside wall of the bowl which encourage the dough to tumble around inside the mixing bowl. This action aids energy transfer and the mechanical development of the dough.

It is common in CBP-compatible mixers to fit some form of mixer headspace control. In its 'classic' form this consisted of a vacuum pump capable of reducing the headspace pressure to 0.5 bar (Pickles, 1968). With the loss of potassium bromate as a permitted oxidising agent in UK and elsewhere, the relationship between headspace atmosphere and ascorbic acid became more critical and the pressure–vacuum mixer was developed in which mixer headspace pressures can be varied sequentially above and below atmospheric pressure (APV Corporation Ltd, 1992). With the 'pressure–vacuum' mixer it is possible to produce a wide range of bread cell structures through adjustment of mixer headspace pressure both above and below atmospheric pressure and in various sequences (Cauvain, 1994, 1995).

The essential features of the pressure–vacuum mixer are as follows:

1   A mixing chamber capable of withstanding up to 4 bar pressure absolute (i.e. 3 bar above atmospheric pressure). The same mixing chamber is capable of withstanding the application of partial vacuum to a level of 0.27 bar.

2   An air inlet orifice beneath the mixing plate in the bottom of the bowl so that air under positive pressure can enter the bowl during dough mixing and that pressure can be maintained using a suitable control valve in the lid.
3   A suitable control mechanism for pre-programming energy input.
4   Suitable mechanisms for the control of pressure inputs and their sequencing. The length of time for sequenced applications of positive pressure or partial vacuum can be varied independently within the total mixing time and is usually adjusted on the basis of energy expenditures rather than time.

In another possible variation of mixer headspace control it is possible with some CBP-compatible mixers to replace the atmospheric headspace gas with different gas mixtures. Most successful has been the application of a mixture of 60% oxygen and 40% nitrogen based on the principles of providing improved ascorbic acid oxidation (Chamberlain, 1979).

Control of the mixing process in the CBP is through the delivery of a defined energy expenditure rather than using mixing time. This is achieved using a watt–hour meter or similar energy-measuring device attached to the mixer. The required energy input for each mix is input in advance of starting the mixing process and the mixer is automatically turned off when the energy has been delivered. In the UK energy levels of around 11 Wh/kg of dough in the mixer are common while, in other parts of the world or with products such as breads in the USA, this may rise to as much as 20 Wh/kg of dough (Tweedy of Burnley Ltd, 1982, Gould, 1998). In the production of US-style breads where fine cell structures and higher energy inputs are required to achieve optimum dough development, CBP-compatible mixers may be fitted with a cooling jacket to maintain control of final dough temperatures (French and Fisher, 1981).

Mixers based on the original Tweedy design and suitable for the production of dough by the CBP include those supplied by the following:

1   APV Baker Ltd (Peterborough, Cambridgeshire, UK).
2   Turkington Industries Ltd (Burnley, Lancashire, UK).
3   ECS Engineering (Christchurch, New Zealand).
4   SAF Yeast Kneady mixer (SAF Yeast Co. Ltd, Mumbai, India).

No doubt there are other locally built Tweedy-type mixers. Tweedy 10 laboratory mixers have also been available in the past, although there have been concerns that these mixers have not always been representative of the dough-mixing effects which might be experienced on a larger scale. More recently Paragon Bakery Systems Ltd (Burnley, Lancashire, UK) have produced a laboratory-scale pressure–vacuum mixer.

Mixers not based on the Tweedy design which have the potential for delivering the required energy in the specified time range have included the following:

1   The Oakes continuous mixer–modifier.
2   The Strahmann continuous dough developer.
3   The Wallace and Tiernan Do-Maker.
4   The Amflow mixer.
5   The Bi-Plex mixer (Anon., 1983).
6   The Uni-Plex mixer (Anon., 1988).
7   VMI Verymix (continuous).
8   Diosna twin spiral.
9   Diosna in-line mixer.

The role of energy

As discussed earlier, it is necessary to 'work' the flour and water mass in order to generate a gluten structure and to form a dough, even if that work is input by hand mixing. The delivery of a pre-determined amount of energy during dough mixing is a basic requirement for the CBP. The total energy required is directly related to the quantity of ingredients in the mixing bowl; hence energy levels are commonly quoted in watt-hours per kilogram of dough (the total ingredient mass) in the mixer. As the level of energy per kilogram of dough in the mixer increases, bread volume increases and with the increase in bread volume comes a reduction in crumb cell size (i.e. the crumb becomes finer), increased cell uniformity and improved crumb softness. The input of energy during mixing causes a considerable temperature rise to occur. For an energy expenditure of 11 Wh/kg the final dough temperature will be about 12–14 °C higher than the weighted average of the initial ingredient temperatures. The contribution that dough temperature plays in bread quality and the practical consequences of temperature rise in the CBP are considered further below.

The role of energy during CBP mixing has yet to be fully explained. In part, it provides a means by which the hydrated protein (the gluten strands) are stretched to form microscopic plates and sheets. As the gluten strands are moved with continued mixing, they continually pass by one another and there is the potential for cross-linking through the action of any oxidants present. The most readily understood bonding mechanism in bread making is the formation of disulphide bonds between protein chains through the disulphide–sulphydryl interchange. In CBP dough mixing, we can consider that the effect of mechanical energy will be to smash any bonds that may have formed. However, Chamberlain (1985) considered that only about 5% of the available energy was required to break the disulphide bonds with the rest being consumed by mixing of the ingredients and the breaking of weaker bonds. An unknown factor will be the proportion of broken disulphide bonds that may subsequently reform and so it is possible that the value given by Chamberlain is an underestimate, although not by a large amount.

Weaker bonds which will be present in dough during mixing include the formation of hydrogen bonds. These will be present in far greater numbers than the disulphide bonds and, although the energy required to break an individual bond will be low, the overall effect on the energy requirement from the large numbers of bonds may be significant. More recently (Tilley *et al.*, 2001) arguments have been put forward for the formation of potentially significant numbers of dityrosine cross-links in dough, including those made by the CBP (Cauvain *et al.*, unpublished observations). The potential contributions that hydrogen and dityrosine bonds make to energy expenditure in the CBP have yet to be quantified but will account for a reasonable proportion of the energy expenditure in CBP mixing.

Kilborn and Tipples (1974) used sheeting rolls and a 'laminating' process to study the 'development' of bread dough and concluded that development could be achieved with only 10–15% of that required with a pin mixer. Such findings imply that the high levels of energy expenditure commonly used in the CBP are unnecessary for dough development. This may be the case but such low energy values still have to be reconciled with the practical observations made when the CBP was launched as a process and indeed have been made ever since, namely that reductions in the practical optimum energy expenditure for a given flour lead to loss of bread volume and changes in bread cell structure, usually a coarsening.

The speed at which mixing is carried out and therefore the rate at which energy is transferred to the dough has an important effect on the rate of transfer of energy to the dough. The original specification for the CBP is that the energy should be delivered in not less than 2 min and not more than 5 min. Typically for 11 W h/kg this requires a mixing speed of the order of 300 rev/min. However, in the case of higher-protein stronger flours which require high energy inputs, e.g. 13 W h/kg or more, there are advantages in increasing the mixing speed so that the energy can be delivered within the critical time frame of 2–5 min. Cauvain (1998b) illustrated the effect of changing mixer speed with two flours and showed that the stronger of the two responded positively to an increase in mixing speed while the weaker of the two showed limited response to changing the rate of work input.

The alternative to mixing at higher speeds is to mix for longer times. While this provides an apparent practical alternative for delivering higher energy levels, it has been shown on numerous occasions that changes in bread quality, some of which are undesirable, do occur. Millar (2003) provided an interesting comparison of breadcrumb derived from mixing two flours, one strong and one weak, at 300 and 100 rev/min. The illustration shows that the strong flour mixed at 300 rev/min gave a fine crumb cell structure which became open when the mixing speed was lowered, even though the total energy delivered remained constant. On the other hand, the weak flour yielded an open crumb cell at 300 rev/min which was little changed when the mixing speed was lowered to 100 rev/min.

The rate at which energy is transferred to the dough during mixing with the CBP is strongly influenced by the mixing-bowl geometry and the relationship between dough mass and mixing-bowl volume. Once the dough has formed in a vertical CBP-type mixer, the relatively high speeds used mean that the impeller picks up the bulk dough and propels it upwards and outwards to impact with the side of the bowl, as discussed above. After impact the dough falls to the bottom of the mixer under the influence of gravity where it is once more picked up by the impeller and propelled to the outside of the bowl. The continual impact of the dough with the bowl walls results in the transfer of energy and contributes to the rise in dough temperature during mixing. The contribution of the impeller to energy input is most likely to be associated with the efficiency with which the dough is picked up and thrown around the bowl.

Vertical CBP-type mixers commonly have a bowl design which incorporates a number of projections on the inside of the mixing bowl (see example in Fig. 5.1). The function of these internal baffles is to assist the return of the dough into the centre of the impeller and, as such, they have a significant effect on the efficiency of the mixing action. Their number and location appear to have a profound effect on the rate of energy transfer to the dough but few data are available to support the efficacy of any particular design.

If the total mass of ingredients (dough) in the bowl changes, then so does its relationship with the mixing bowl volume and its interaction with the mixer impeller. This, in turn, leads to changes in the rate at which energy is transferred to different masses of dough. The practical consequence of changing dough mass in the mixing bowl is that the mixing time required to deliver a given energy will vary. Thus, if a fixed mixing time is used, then any variations in dough mass in the mixing bowl will result in variations in energy expenditure and, in turn, dough development and final dough temperature. While the degrees of variability may be small, such variations are contrary to one of the main requirements of the baker, namely that bread quality should be as consistent as possible from batch to batch. Variations in final dough temperature can be particularly damaging in a bakery because of the impact on yeast activity, as discussed below. The consistency of the dough will also affect the rate at which energy is transferred to it during mixing. Energy transfer is based in part on the resistance of the dough to the movement of the impeller with stiff doughs having greater resistance than soft doughs. Thus, mixing times to a given energy input will be shorter with stiff (low-added-water) doughs than with soft doughs.

The above comments on dough mass and consistency and the need to reduce batch-to-batch variability explain to a large extent why it is important to work to a fixed energy in the CBP. The calculation is simple and based on the total weight of the ingredients (dough) mass multiplied by the required energy expenditure for a given flour and desired bread quality. If the mixer is fitted with a watt–hour meter of some form, then it is a simple exercise to input the required energy expenditure. The failure to understand the importance of using a measured energy value rather than a fixed mixing time by many users of the CBP was one of the reasons for its apparent rejection in some practical bakery environments (see also Chapter 10).

The role of temperature

Mechanical work is intimately linked with heat as anyone who has mixed dough by hand will confirm. In CBP-type mixers the dough is subjected to intense mechanical work and this is translated into heat and manifest in the temperature rise which occurs. While it is not possible to divorce the effect of work from heat rise, there is the question as to what dough temperature should be used in the CBP and what, if any, is the link with dough development. The link is so strong between temperature rise and energy that it is possible to estimate energy expenditure during mixing based on ingredient and dough temperature data (Chamberlain *et al.*, 1967). The calculation provides a convenient and practical means of checking the performance of mixers and associated watt–hour controllers in the CBP (or in any other breadmaking process).

The most commonly quoted temperature range for the CBP is between 28 and 30 °C. In part, this is to provide a degree of control over the rate of gas production by the yeast and in part to influence dough rheology for processing. The common CBP temperature range is some 4–6 °C higher than would be seen as 'ideal' for bulk fermented doughs where control of dough temperature is more critical since fermentation time is a key element of the dough development process. Even though the CBP uses higher yeast levels than commonly used in bulk fermentation, bakers' yeast experiences a 'lag time' during which little carbon dioxide is being produced. Typically this lag time is around 10–15 min in length which is sufficient time for most CBP doughs to have been divided and moulded at least once.

In most cases, the control of final dough temperature to the CBP 'norm' requires the provision of chilled water. In those parts of the world where ambient bakery temperatures can be very high or where high energy levels are required to develop the dough fully, it can be difficult to achieve the target of 28 °C and 30 °C. Two options are available: one is the provision of a cooling jacket for the mixer and the other is to run doughs at temperatures above 30 °C. Water- or glycol-chilled mixers have been used with CBP doughs with some success (French and Fisher, 1981) but they have not become common.

One of the problems associated with higher dough temperatures ex-mixer in the CBP is the change in dough rheology which occurs. In common with many materials, higher temperatures reduce the viscosity of CBP doughs, i.e. the doughs become softer. The practical consequence of this change is that dough dividing and especially first moulding become more difficult to achieve. In many bakeries the adjustment of equipment to handle the softer dough results in smearing of the dough on to the equipment surfaces and the dough is described as 'sticky'. The practical solution is to reduce the added water level in the dough with subsequent loss of yield.

In view of the undesirable changes in dough rheology with higher temperatures the pragmatic approach in the bakery is to restrict the final dough temperature to a value which allows the doughs to be processed with existing equipment. However, if alternative dough-processing methods are available, then there is no real reason why dough temperatures higher than 30 °C cannot be used with CBP doughs. Indeed, in India, final dough temperatures for CBP doughs may reach 36 °C or higher but with modified mechanical handling or hand moulding the dough pieces can still be processed for pan bread production.

There are positive benefits associated with the higher final dough temperatures achieved in the CBP. In addition to greater potential yeast activity, there will be increased dough oxidation through enhanced ascorbic acid (and other oxidant) activity. The activity of enzymes present in the flour and the improver will also be enhanced. If bread made from doughs prepared with the same work input but mixed to final dough temperatures of 30 °C and 21 °C are compared, that made at 21 °C lacks volume by comparison with that mixed at 30 °C. If the dough pieces are proved to the same height before entering the oven, the difference in volume comes from differences in oven spring which reflect differences in dough gas retention. Collins and Haley (1992) also found that to be the case in the production of dough for freezing. In addition to the improved gas retention, doughs with higher final temperatures yield benefits during dough processing. A comparison of the impact of final dough temperature is summarised in Table 5.1.

Mixing and dough consistency

The need to optimise the level of water added for dough mixing is common for all types of breadmaking process. If there is too little water, the dough is 'tight' and difficult to process; if there is too much water, the dough becomes soft and sticky and difficult to process. Both water deficiency and water excess lead to quality problems. In the bakery the need for a consistent dough rheology is important in delivering a consistent performance on the plant and consistent final product quality.

In the CBP, variations in dough rheology during the mixing cycle may yield variations in mixing times. This occurs because the resistance of the dough to the action of the impeller blades will vary with dough consistency. Low water and therefore high dough viscosity yield a dough with greater resistance and, in consequence, mixing times may decrease. The reverse is

Table 5.1 The impact of final dough temperature on dough and bread quality

Recipe, dough or bread property	Impact of the following dough temperatures ex-mixer	
	30 °C	21 °C
Yeast level	Can be reduced to maintain given proof time	Must be increased to maintain given proof time
Ascorbic acid	Increased activity potential Lower levels might be used	Reduced activity potential
Fat	Higher solid fat content required	Lower solid fat content may be used
Water level	May need to be reduced	Higher rates of addition can be used
Dough rheology	Less resistance to deformation and dough damage during moulding	Greater resistance to deformation and dough damage during moulding (see water level)
First or intermediate proof	Can be relatively short with 'relaxed' dough	May need to be increased with 'tight' dough
Final proof	More uniform because dough piece warms more rapidly	May require longer time in order to achieve more uniform heat distribution
Baking	More uniform expansion of dough in oven and shorter baking time may be achieved	Less uniform expansion of dough and longer baking time required

true for high-water doughs. In order to ensure consistent dough mixing and dough development in the CBP it is therefore necessary to mix to a defined energy input rather than to time. Thus, the watt–hour meter is a key factor in delivering consistent dough and product quality.

With open-bowl mixers and the longer mixing times typically associated with their use, adjustment of dough consistency through the addition of extra water is possible. Thus, a common practice in bakeries with respect to optimising water addition is to start with a slightly water-deficient dough and to assess the rheological characteristics of the dough part of the way through the mixing cycle. Typically the assessment is carried out in the hands of an expert baker. This approach allows for further additions as required once the dough has formed in order to optimise dough consistency ex-mixer. In the closed bowl types of mixer commonly used in the CBP, however, access to the dough part of the way through the mixing cycle is not feasible.

The lack of ready access to the dough and the relatively short mixing cycle used in the CBP do not mean that adjustment of dough consistency is not possible. Russell Eggitt (1975) discussed the development of an automatic dough consistency controller fitted to Tweedy mixers. The concept was based on the measurement of mixer power consumption with increasing mixing time in the CBP. The general form of the traces was a rapid increase to a maximum height and after a short time at that maximum height, followed by a gradual tailing off until mixing was stopped. Three dough consistencies were used by Russell Eggitt, namely water deficient, optimum and water excess, and in all cases the forms of the curve were the same. Russell Eggitt (1975) showed that it was possible to adjust the consistency of a water-deficient dough through the addition of extra water at about halfway through the mixing cycle. To use this approach effectively it was necessary to start with a water addition that was lower than expected (thus mimicking common craft practice) and for the level of additional water to be small; otherwise there was risk of not having sufficient mixing time to ensure uniform dispersion of large amounts of additional water. In practice the need for automatic dough consistency control has largely been eliminated by improvements in the control of load cells delivering flour to the mixer. Now the actual rather than the nominal weight of flour delivered to the mixer is known and the level of water addition adjusted accordingly.

Cell creation during mixing

The production of a defined cellular structure in bread and other fermented products depends entirely on the creation of small bubbles of gas in the dough during mixing and their retention during subsequent processing. Since most dough mixing is carried out in air, the principal gases that are incorporated are nitrogen and oxygen, in the approximate ratio of 4:1 by volume. The contribution of oxygen to the dough oxidation processes which take place in the dough have been discussed above. The residence time for oxygen in normal dough mixing is relatively short because it is quickly scavenged by the yeast cells which are present (Chamberlain, 1979).

With the removal of oxygen from the dough the only gas that remains entrapped is nitrogen and this plays a major role by providing bubble nuclei. After mixing has been completed, the only 'new' gas which becomes available is the carbon dioxide gas generated by the yeast fermentation. Carbon dioxide gas has a high solubility relative to other gases and in bread dough cannot form gas bubbles on its own (Baker and Mize, 1941). In the early stages of fermentation the carbon dioxide dissolves in the aqueous phase in the dough until saturation is reached after which the gas diffuses into the nitrogen gas bubble nuclei which then begin to expand.

The critical role that the creation of gas bubble nuclei plays in the CBP is illustrated in Fig. 5.2 using doughs containing a mixed oxidation system of ascorbic acid and potassium bromate. While mixing doughs in an atmosphere of 100% oxygen produces doughs with gas levels ex-mixer of around 4%, the disappearance of the oxygen, as shown by the fall in gas volumes

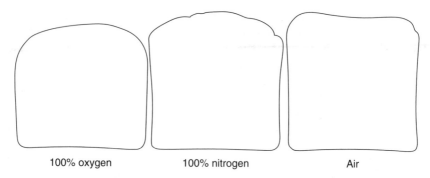

Fig. 5.2   Effects of changing the gas composition during dough mixing.

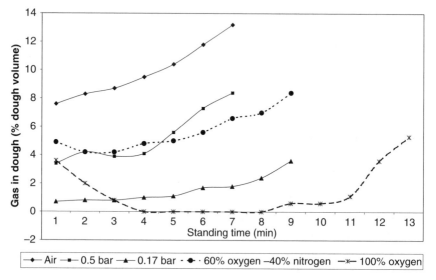

Fig. 5.3   Changes in the dough gas volume with standing time after mixing.

with time (Fig. 5.3), considerably reduces the gas bubble nuclei potential in the dough. The bread volume from dough mixed in 100% oxygen is greater than that of a dough mixed under high vacuum (less than 0.15 bar) because there is some ascorbic acid-assisted oxidation but the lack of gas bubble nuclei leads to the coarse cell structure seen in the bread. Doughs mixed under a mixer headspace gas consisting of 100% nitrogen yield further improvements in bread volume which come from the addition of potassium bromate. The absence of oxygen for the ascorbic acid oxidation was respon- sible for the smaller volume compared with that of the dough mixed in air. However, the presence of nitrogen to provide the necessary gas bubble

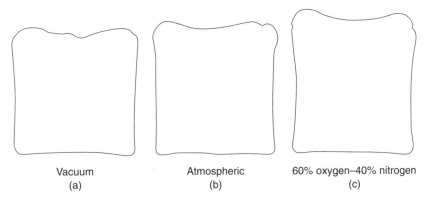

Vacuum   Atmospheric   60% oxygen–40% nitrogen
(a)   (b)   (c)

Fig. 5.4 Comparison of bread made with different mixer conditions: (a) partial vacuum: (b) atmospheric pressure; (c) oxygen-enriched gas at atmospheric pressure.

nuclei led to a significant improvement in crumb cell structure. The application of a significant level of vacuum (about 0.17 bar absolute pressure) removes most of the air from the dough, a fact confirmed by Collins (1986) who recorded gas volumes in such doughs as being 0.5% ex-mixer (Fig. 5.3). The impact of reducing the mixing headspace pressure is shown in Fig. 5.4.

The total volume of gas occluded during dough mixing is known to vary from mixer to mixer (Marsh, 1998), even those which are CBP compatible. Thus the quantity of oxygen which is available will also vary. This accounts in part for the differences seen between the bread qualities for different mixers, even when they can be set to deliver similar levels of total energy in the 2–5 min range specified for the CBP. The other significant contribution to differences between the bread qualities for different mixers lies in the sizes and numbers of the gas bubble nuclei in the dough. In no-time dough-making processes, such as the CBP, the 'gas bubble populations' created during mixing form the basis of the final product cell structure and other quality attributes. After the dough leaves the mixer, no 'new' bubbles are created although significant changes take place in their numbers and dimensions during processing. As a result of the yeast fermentation processes, bubbles in the dough may grow by inflation with carbon dioxide gas or they may shrink. The former change is likely to be the most common, although there is evidence that the latter change may be more common than previously considered (Whitworth and Alava, 2002). In the CBP it is still essentially the case that the bubble structure created in the mixer is essentially the structure that will be expanded during proof and set in the oven. Thus control of gas bubble creation during mixing is an essential feature in the creation of a particular structure in the final product.

The numbers and sizes of gas bubbles in the dough at the end of mixing are strongly influenced by the mechanism of dough formation and the mixing conditions in a particular machine. Recent work to measure bubble distributions in CBP bread doughs (Cauvain *et al.*, 1999) has confirmed that different mixing machines do yield different bubble sizes, numbers and distributions. However, in one CBP-compatible mixing machine, variation of impeller design had only a small effect on the gas bubble population (Whitworth and Alava, 1998). This lack of difference in the characteristics of the dough bubble populations was confirmed by the absence of discernible differences in the subsequent bread cell structures.

In addition to the mechanical effect on the creation of bubble structures in the dough, many ingredients contribute to the gas bubble populations which form. Whitworth and Alava (2002) have shown that changes in flour type resulted in different gas bubble populations in the dough after mixing. The differences remain in the dough during processing and proof and are seen later as differences in breadcrumb structure.

The precise mechanism by which gas bubbles are trapped in dough during CBP mixing is not fully understood. Air is clearly trapped with the bulk of the flour when it enters the bakery and, as gluten development proceeds during mixing, it is likely that the gluten films will occlude some of this air as gas bubbles in the dough structure. The other source of air during CBP mixing is that contained in the mixer headspace some of which may become trapped during the folding and stretching of the dough during mixing. The latter action is most likely to occur when the dough is deflected by the baffles on the bowl as the impeller sweeps by. The frictional forces of the baffles will retard a portion of the dough while the movement of the impeller will try to pull it into the centre of the bowl. Thus, for short periods of time, portions of dough will be stretched and, when these portions are folded by continued mixing, there is the opportunity for the entrapment of air.

While a reasonable explanation can be given for the mechanical entrapment of air into CBP dough, the mechanism by which the size of the gas bubbles is controlled is more speculative. Given that different flours yield different gas bubble populations by the end of mixing, then clearly the inherent properties of the gluten-forming proteins in the flour are a key component of the mechanism. It is also known that some ingredients, e.g. fat and emulsifers, also aid in the formation of gas bubble populations.

Gas volumes occluded in CBP dough during mixing increase during the early stages of mixing but after a period of time there are no further increases (Campbell, 2003). At some time during the mixing cycle the point is reached at which the volume of gas entrained during mixing equals the volume of gas lost because of the mechanical actions in the mixer. The latter action is referred to as disentrainment. The point at which equilibrium is reached varies with mixer type and in the case of some mixers may never be reached before the end of the mixing cycle.

Mixer headspace pressure control

The survival and growth of particular gas bubbles are direct functions of their initial size and as such are influenced by the many factors that contribute to gas bubble formation and stability. Such controlling factors are common to most breadmaking processes. However, in the CBP a further dimension is added by the ability to control directly the mixer headspace pressure. Initially this would have been through the application of partial vacuum to 0.5 bar but with the development of pressure–vacuum CBP-compatible mixers a wider range of pressures and more combinations of effects are possible.

Mixing dough at pressures above atmospheric creates a more open bread cell structure, i.e. a larger average cell size (Cauvain and Little, 2000). However, differences between loaves made from doughs mixed at pressures of 3 and 4 bar are small. Accompanying the change in cell structure there is a small increase in bread volume as the mixing pressure increases. As would be expected, the bread cell structure became progressively finer (i.e. a smaller average cell size) and the volume decreased as the mixer pressure progressively decreased below 1.0 bar. Whitworth and Alava (1998) presented bubble population data for doughs mixed at a pressure of 1.5 bar, which showed a greater range of bubble sizes, ranging up to 1400 μm compared with sizes up to 75 μm when mixed at atmospheric pressure. Not all the large gas bubbles would be expected to remain as bread cell structure in the final bread, but the 'average' bubble size would be greater and this is reflected in the greater 'average' cell size in the breadcrumb.

The ability of the pressure–vacuum mixer to be programmed to run a sequence of mixer pressures enables increased control of both ascorbic-acid-assisted oxidation and dough gas bubble populations. A typical sequence for sandwich-style bread is run first above atmospheric pressure during mixing (to optimise ascorbic-acid-assisted oxidation), followed by a period below atmospheric pressure to reduce gas bubbles to a suitable small size in the second part. Data recorded by Whitworth and Alava (1998) confirm that the application of partial vacuum during the second part of the mixing cycle is able to reduce both the gas volumes and the overall size range of the gas bubble populations. Figure 5.5 shows a comparison of bread created with sequential application of positive and negative pressures and bread made from dough mixed under partial vacuum. With changes in mixer headspace pressure, changes in the total volume of gas occluded occur. The volume of gas occluded increases as the mixer headspace pressure increases and decreases as the headspace pressure decreases.

Changes in the volume of gas present in dough at the end of mixing lead to differences in dough density and in dough rheology. Cauvain and Little (1999) used large-scale deformation testing to study the effects of mixing and processing conditions on CBP dough. They found that the application

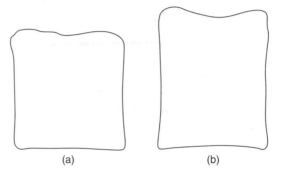

Fig. 5.5   Comparison of bread made from dough mixed under (a) a partial vacuum
and (b) a positive pressure and a partial vacuum in sequence.

of partial vacuum, whether delayed or not, increased dough resistance to
deformation. Doughs mixed under such conditions contrasted those mixed
at atmospheric or under positive pressure. Cauvain and Little (2000) also
linked such changes to gas volumes in the dough, showing that low gas
volumes in CBP dough made them more resistant to deformation. These
findings are consistent with the practical observations that CBP doughs
mixed under partial vacuum feel 'firmer' and therefore require more water
to be added in order to achieve a similar dough rheology to that of dough
from other breadmaking processes.

As discussed above, the availability of oxygen in the CBP is particularly
important when ascorbic acid is the sole oxidising agent. An alternative to
increasing air occlusion in the mixer headspace is to increase the concentration of oxygen in the mixer headspace. Chamberlain and Collins (1979)
studied the role of nitrogen and oxygen and found that improved oxidation, as shown by increased bread volume and finer crumb cell structure,
could be achieved with the CBP using a mixture of 60% oxygen and 40%
nitrogen, the latter acting as a nucleating gas for the carbon dioxide. The
effects of increasing oxygen concentration on loaf height are illustrated in
Fig. 5.4. Breadcrumb cell structure changed from being coarse and open
at 100% oxygen to gradually becoming finer until 60% oxygen and 40%
nitrogen, before again becoming coarser at 100% nitrogen. The improvement in cell structure (i.e. finer) which occurred with the optimisation of
oxygen concentration (60%) was of the same order as would be observed
when applying a partial vacuum during dough mixing. The coarse cell structure obtained with 100% oxygen or 100% nitrogen was comparable with
that seen when very high levels of vacuum are used during CBP dough
mixing.

In the pressure–vacuum mixer, greater ascorbic acid-assisted oxidation
can be achieved by adjusting the ratio of mixing times under positive pres-

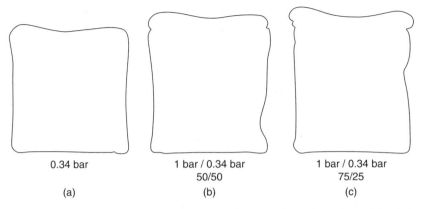

0.34 bar

(a)

1 bar / 0.34 bar
50/50

(b)

1 bar / 0.34 bar
75/25

(c)

Fig. 5.6   Effect of changing the pressure-to-vacuum ratio during mixing: (a) mixing under partial vacuum; (b) mixing under 50/50 pressure/partial vacuum; (c) mixing under 75/25 pressure/partial vacuum.

sure to that under negative pressure (partial vacuum). Commonly, mixing ratios are 50/50, i.e. 50% of the energy expenditure under positive pressure, followed by the remaining 50% under partial vacuum. The effect of increasing the time spent under positive pressure is illustrated in Fig. 5.6, which shows that, when the ratio was changed to 75/25, there was an increase in bread volume. It is important in such circumstances that the negative pressure is quickly achieved and held for some time before the mixing cycle is completed. Typically, a period of some 15–20 s under partial vacuum is required. The ability of a given mixer to achieve this result will depend on many factors, not least the operating efficiency of the vacuum pump.

Brews, sponges and sours

The development of flavour in fermented products is derived from the ingredients, the formulation and the processing methods which are used in their manufacture. During the dough fermentation process, new flavour products are generated within the dough. The intensity of those flavours and particular flavour 'notes' which are developed change with increasing fermentation time. The most commonly observed flavour changes are those associated with the development of acid flavours from microbial activity in the dough which are readily detected in the flavour of the breadcrumb. Not all this flavour activity will come from the addition of bakers' yeast; some will come from wild yeasts and bacteria, especially lactic acid bacteria, which are present naturally in the flour. Usually several hours of fermentation are required before there are significant changes to the flavour profile of the breadcrumb. Bread flavour may be enhanced using a 'pre-ferment', 'brew' or 'sponge' based on a portion of the ingredients which are later mixed with the remaining ingredients to form the dough for final processing.

The various forms of pre-ferments, sponges and brews (Collyer, 1967; Thompson, 1983) may be used with the CBP to modify bread flavour. The formulation of the sponge and the length of fermentation time used can vary according to the flavour profile required. An example recipe for a sponge suitable for the CBP is given in Chapter 9. The pre-ferment, sponge or brew component should be omitted from the energy calculations at the dough stage. It is generally not recommended to include ascorbic acid in the pre-ferment or sponge component because it is likely to work as a reducing agent in the anaerobic environment that forms in the mixture. The use of a sponge with CBP allows a small reduction in the overall yeast level. There will be slightly more gas in CBP doughs made with a sponge but the overall levels will still be much lower than from other dough-making processes and should cause no problems with divider weight control. Pre-ferments may be used with dough mixed at atmospheric pressure, partial vacuum or positive pressures, or combination pressures, as with pressure–vacuum mixers.

Bread flavour may also be modified through the addition of dried sours. These are formed by controlled fermentation of wheat and other cereal flours, most commonly rye (Stolz and Bocker, 1996). A number of different micro-organisms may be used in the fermentation process. After the prescribed fermentation period the soft dough-like material is dried to yield a dry powder. Such powders can be added directly at the mixer together with the other dry ingredients. The quantity of dried sour used should be taken into account when making the energy calculations.

6

Processing Chorleywood Bread Process doughs

Modification of dough bubble structure

In all no-time doughs the gas bubbles in the dough are at their smallest size when the dough leaves the mixer. Generally the size of the gas bubbles increases during dough processing to bread. Some small bubbles may shrink but, in doing so, they commonly disappear. Coalescence of gas bubbles occurs as fermentation progresses so that the number of gas cells left in the bread is a small fraction of the number of gas bubbles created in the mixer. Coalescence readily occurs in proof and in the early stages of baking when most of the carbon dioxide gas is being produced. It may therefore be seen as a 'natural' and acceptable phenomenon in the production of fermented products. Typically the sizes of gas bubbles in dough from the mixer lie in the range 0.05–0.20 mm, and the cells that are seen in the product crumb will lie in the range 0.2–500 mm. The precise sizes will depend on many of the factors discussed earlier.

The coalescence of gas bubbles discussed above may well be considered to be intentional and to the benefit of product quality but there are occasions when coalescence is unintentional and to the detriment of product quality. Such unintentional coalescence is commonly considered under the heading of 'dough damage', i.e. disruption of the 'normal' gas bubble population of the dough. As such, dough damage is associated with the breakdown of the gluten membranes which are a key part of the gas-bubble-stabilising mechanism in fermented doughs (i.e. they keep individual bubbles apart while they grow during fermentation). Gluten breakdown arises from the complex interaction between the dough rheology and the action of the mechanical equipment used in the processing of doughs. Damage to the dough gas bubble structure can occur when doughs are processed by hand but generally the sensitivity of a skilled baker's hands adjusts pressures, stresses and strains on the dough to reduce the potential for damage. No equipment currently used in the manufacture of fermented products can achieve the same degree of responsiveness to variation in dough rheology, whether deliberate or accidental.

There are planned modifications to gas bubble populations in dough in addition to those seen in the prover and the oven. Most of these are associated with dough processing from the divider to the time of entering the prover and are discussed in more detail below. Such modifications are designed to be beneficial but they also have significant potential to be detrimental to final product quality. This is especially true for CBP and other no-time doughs and this situation arises in part because much of the dough-processing equipment still uses designs that are more applicable to the rheology of bulk fermented doughs which contain higher levels of gas than typically seen with CBP doughs. These different gas levels and the longer processing times contribute to the significant differences between the dough rheologies of different dough-making processes.

Dividing

After mixing, the bulk dough is divided to generate the shape and size of product required. Dough is most generally divided volumetrically with portions of a given size delivered either by filling a chamber with dough and cutting off the excess (piston dividing) or by pushing the dough through an orifice at a fixed rate and cutting billets from the end at regular intervals (extrusion dividing). In either case the accuracy of the system depends on the homogeneity of the dough.

Typically CBP doughs contain little gas at the end of mixing, especially if partial vacuum has been applied towards the end of the mixing cycle. In the first few minutes after mixing, yeast activity is low and there is little change in dough density. Thus, if the dough is processed quickly after mixing, there will be little need for excessive divider adjustment to maintain product weight control. Since CBP doughs contain more yeast and are generally made to higher dough temperatures, it is important to minimise the time which a dough spends in the divider hopper. Thus the mixing cycle must be matched as closely as possible to the divider speed. If very high production throughput rates are required, then it may be necessary to install more than one mixer to feed the divider.

The dough may be worked considerably during the dividing process and this can cause mechanical damage to the dough's gluten structure. Working the dough at this stage should not be compared with mechanical development. The difference is that the mechanical work done during mixing is uniformly distributed throughout the dough structure, whereas mechanical work during such dough transfer systems is not uniformly distributed and confers different changes in dough properties in different areas of the dough mass. Degassing of CBP doughs should not normally occur because gas levels are already low. However, any attempt to do so should be discouraged. If carried out with a minimum of stress on the dough, little change occurs to gas bubble populations during dividing (Whitworth and Alava, 1999).

Rounding and first moulding

After dividing, the individual dough pieces are commonly worked in some way to change the rough shape coming from the equipment. The most common shaping is by rounding, an action which mimics that carried out by hand in the craft bakery. The action of mechanical rounding places the dough under stress and strain and may lead to damage to the existing gas bubble structure in the dough. Some breadmaking processes require the rounder to have a degassing effect; however, as noted earlier, CBP doughs have little gas in them when they first leave the mixer and so degassing is not normally required. Indeed Cauvain (1997) used X-ray computerised tomography to show that quite large-sized air pockets could remain trapped in CBP dough pieces coming from the rounder but that such large air pockets did not necessarily survive through to the baked product. In part, this is because the gas bubbles would be too large to pass through the sheeting rolls of the final moulder without being changed (see discussion below).

Cauvain and Little (1999) studied the changes that occurred in CBP dough rheology at the various processing stages and the influence of different mixing atmospheres. Before rounding, the surface of the dough pieces may exhibit a 'tackiness' or 'stickiness' which quickly disappears, especially if subjected to the passage of an air stream or the application of a dusting flour. Any problems with dough stickiness during first moulding probably occur because the dough pieces are subjected to shearing forces in the first moulder. Cauvain and Little (2001) studied the contribution of the first moulder to CBP bread quality. They came to the conclusion that an important function of the first moulder was to establish a uniform shape for presentation to the final moulder and the elimination of the rounding had no adverse effect on CBP sandwich bread quality provided that dough pieces were delivered uniformly to the final moulder (see discussion below).

Intermediate or first proving

In most modern dough make-up processes, intermediate or first proof is used as a period of rest between the work carried out by dividing and rounding and before final shaping. The length of time chosen for this process is related to the dough rheology from the mixer and rounder and that required for final moulding. Generally speaking, doughs arriving at the final moulder are expected to have little resistance to deformation, limited elasticity and good extensibility. Ingredients, dough formulations and processing conditions all affect the dough rheology and the changes that occur as it rests. The longer that dough rests, the greater will be the changes in dough rheology, even in the absence of yeast. Commonly the collective changes are referred to as 'relaxation' of the dough. The extent of the relaxation depends on the length of time involved, the dough temperature and many other formulation and recipe factors.

In intermediate proof the density of the dough usually decreases as the yeast begins to produce carbon dioxide gas. The speed at which the dough density reduces depends in part on the dough temperature and yeast level used in the dough. Collins (1986) studied the changes in CBP dough density with time after completion of mixing and compared the effects of different mixer atmospheres. He found that, the lower the mixer headspace pressure (greater vacuum), the lower are the gas levels in the dough and the slower is the initial decrease in dough density (see Fig. 5.3.). The size of the gas bubbles in the dough increases as the carbon dioxide gas produced by the yeast diffuses into the gas bubbles present (Whitworth and Alava, 1999). The main effect is an overall increase in size; while there appears to be an increase in numbers of measured gas bubbles, this is probably the result of very small gas bubbles (less than $20\mu m$, the lower limit of the measuring techniques used) which were inflated to reach a measurable size.

In the CBP the degree of expansion is relatively small because intermediate proof times will be short. When first introduced, a 10 min intermediate proof period was the norm for CBP doughs but typically this time will now be less than 6 min and may be as low as 2 min for the manufacture of fine-structured bread, e.g. sandwich-type bread. Some bread types, e.g. baguettes and hearth breads, require an open cell structure for their special character. When using the CBP the creation of the required open structure may be achieved by extending the first proof to encourage gas production. However, a longer first proof is not the only requirement since it is also necessary to ensure that the large gas bubbles so created are not subsequently lost during final moulding. If the appropriate conditions are met, then open crumb cell structures may be readily achieved. Collins (1978) showed that the high specific volume, open cell structure and crisp crust which characterise baguette and pain Parisien could be achieved with suitable processing of CBP doughs (see also Chapter 9).

The complete elimination of first proof in no-time dough-making processes, such as the CBP, causes changes in dough rheology and may lead to a reduction in loaf volume. There is the potential for the lack of relaxation of the dough to lead to an increase in damage to the gas bubble structure in the dough in the final moulding stages which follow. The latter is commonly seen as dark patches of coarse cell structure and dense crumb (Cauvain and Little, 2001). This is especially the case when ascorbic acid is the only oxidant in the recipe because the resistance of the dough to deformation is increased (Cauvain *et al.*, 1992).

Final moulding

The key functions of the final moulder may be considered as shaping the dough to fit the product concept (e.g. shaped to fit a particular pan size or traditional shape) and to reorient the bubble structure. It is essential for optimum product quality that the dough piece is presented centrally to the rollers and is maintained centrally throughout the moulding process (Collins, 1993; Cauvain and Collins, 1995). In many final moulders, the passage of the roughly round dough piece through sets of parallel rolls moving at a high speed is used to reduce the thickness of the dough piece. The gap between successive pairs of rolls decreases and on leaving the last set of rolls the dough piece has an ellipsoid shape. The leading edge of the ellipse created by this process is trapped underneath a static or contra-drive chain which creates a 'Swiss roll' of dough. Compression and further shaping of the Swiss roll are undertaken to give a uniform cylinder of dough. This is achieved by compressing the dough piece underneath a pressure board while it is still being moved along the length of the moulder by the action of a moving belt before it is finally discharged to be placed in a pan or on a tray.

The pressures on the dough during the different stages of this complicated process are significant. The sheeting process acts in part as a degassing mechanism. While CBP doughs tend to have much lower gas levels than dough from many other breadmaking processes, there is still the potential for some degassing to occur. The gap between the sets of rolls is critical in this respect. First the finite gap limits the maximum size of gas bubbles which may survive in the dough. This fact accounts for the non-survival of the large gas bubbles which Cauvain (1997) observed using X-ray computerised tomography scans. However, it must be recognised that the observance of the moulding spiral in computerised tomography scans is entirely due to a slight degassing of the surfaces of the dough sheet passing through the rolls of the final moulder. As the two slightly degassed surfaces are brought together in the curling process, they form a denser layer which has a different X-ray absorbance from the surrounding dough and this shows as a white line. In addition to the slight degassing that occurs when the dough pieces pass through the sheeting rolls, there is an elongation of gas bubbles in the direction of sheeting. Again X-ray tomography has been used to show this very effectively (Cauvain, 1997; Whitworth and Alava, 1999).

Elongation of gas bubbles in the dough sheet occurs because of the viscoelastic properties of the dough and the extent to which it is retained in the final bread structure depends in part on the rheological properties of the dough. The elongation is most likely to occur with the larger gas bubbles located nearer to the surface of the dough during sheeting. It is unlikely that the pressures applied during sheeting will affect the smaller gas bubbles located in the centre of the dough. Nevertheless the elongation of gas bubbles does affect final bread quality because when baked into the bread they tend to be shallower than other surrounding gas bubbles and, since they will cast less shadow in the cut bread surface, they will make the product appear whiter.

For some bread types a further modification of the dough piece takes place at the end of the final moulder or after the dough has left the final moulder. A common form of modification is 'four-piecing' in which the dough cylinder is cut into four equal lengths (each equal to just less than the tin width) and turned through 90° to lie side by side across the tin. This technique reorients the cell structure in the final breadcrumb and is common in the manufacture of sandwich bread using the CBP. Cross-graining, as commonly carried out in the USA, performs part of this function function by turning the dough sheet through 90° as it passes through the final moulder so that a wider elliptical sheet is presented to the curling chain and moulding board. However, cross-graining is very dependent on the rheological properties of the dough and tends to be a more random process than four-piecing.

Given the above description of final moulding operations it is clear that in order to achieve the necessary product quality the dough must have appropriate rheology (i.e. have a low resistance to deformation). This is particularly true when starting from a rounded dough ball. If the dough rheology or the settings of the final moulder are not optimised, then there is the potential for damage to the relatively delicate gas bubble structures in the dough. Cauvain and Little (2001) illustrated the potential for loss of CBP four-piece bread quality through inappropriate final moulder settings. They showed that the areas of damage typically manifesting themselves are dark streaks, swirls and patches in the cross-section of bread slices (see Chapter 7) and could arise from excessive pressure during dough sheeting or passage under the pressure board of the final moulder. They also showed the importance of the correct setting of the guide bars used on the final moulder.

A further problem which may be encountered during dough sheeting is the rupture of gas-stabilising films and the subsequent coalescence of two gas bubbles to form one of larger size. Such larger-sized bubbles have lower internal pressure and carbon dioxide gas may preferentially migrate to such bubbles causing them grow even larger. Such damage to dough bubble structures is thought to be a major factor in the formation of large unwanted holes in breadcrumb (Cauvain and Young, 2000).

Additional benefits of the sheeting process used with CBP dough are improvements in the physical strength of the breadcrumb which enables it to withstand the physical effects of mechanical slicing and buttering. This strengthening of the crumb arises because the dough-sheeting operation tends to align the gluten network in one direction in the dough. After being placed in a pan the upward expansion of the dough also encourages alignment in one direction.

Final proof

Proving remains a key fermentation step in all breadmaking processes, including the CBP. One of the main reactions during proof is the progressive conversion of the starch from the flour into dextrins and sugars by the action of the amylase enzymes, whether naturally occurring or added in the formulation (see Chapter 4). Yeast feeds on the available sugars to produce carbon dioxide and alcohol, as described above. The carbon dioxide diffuses into the gas bubbles in the dough, causing them to grow and the dough to expand. Progressively the size of the gas bubbles increases (Whitworth and Alava, 1999). If the dough is confined in a tin, many of the gas bubbles are elongated in the direction of movement of the dough, i.e. upwards. The frictional forces between the dough and the tin (even when greased) slow down the movement of the edges of the dough piece and so most of the dough expansion occurs in the middle. X-ray tomography has shown that dough expansion in the pan can be so uneven that a point which starts at the centre at the beginning of proof may end up about three-quarters of the way up the dough piece at the end of proof (Whitworth and Alava, 1999). In the case of free-standing (oven-bottom or hearth) breads or those baked in a sling, e.g. baguettes, the expansion of the dough is less confined and so gas bubbles may expand in any direction. The degree and direction of the expansion are largely controlled by the rheological properties of the dough, and in particular by the ability of the gluten network within it to stretch.

Changes occurring during the proving CBP dough have been the subject of a number of studies. Of particular interest has been the formation of larger holes in the dough and their subsequent survival or otherwise in the oven. Especially useful in these studies has been the application of X-ray tomography. Using this technique it has been possible to show the important relationship between damage to the dough piece surface during the sheeting of dough pieces during final moulding (Cauvain, 1997). The slight degassing of the dough yields a slightly denser dough and the proximity of small trapped air pockets in the dough during moulding, especially curling, provides areas of lower pressure into which carbon dioxide can diffuse. Further expansion of the large gas bubbles leads to a further lowering in internal pressure with an increasing likelihood of greater expansion. Once baked, these larger gas bubbles tend to have smooth internal surfaces with no evidence for gluten stranding across the void.

Baking

After proof the dough prepared by any breadmaking method, including the CBP, must be heat set, i.e. baked. Structural stability in the dough piece in the oven is maintained by the expansion of the trapped gases, first carbon dioxide, and later steam and natural thermal expansion of any trapped gas. The ultimate process in baking is one of conversion of a foam in the dough state to one of a sponge in the baked loaf. In the former case the gas bubbles are discrete and separated from one another by delicate gluten films; in the latter the cells of the crumb structure are open and interconnected to one another. The process of foam-to-sponge conversion is crucial to the formation of a final product with the desired appearance and eating qualities. The moment of setting during the baking process depends on many factors, including the rate of heat input and the formulation used. Cauvain and Chamberlain (1988) showed that the bread-improving effect of fungal *alpha*-amylase was partly due to a delay in the time taken for the dough piece to reach maximum height in the oven, this being taken as the time required for the foam-to-sponge conversion to be achieved.

As the dough warms in the oven, a complex progression of physical, chemical and biochemical changes occurs. When the dough piece first enters the oven, its surface temperature is that of the prover, commonly 40–43 °C, while the centre temperature lags a few degrees Celsius behind. There is an initial burst of yeast activity in the first few minutes of baking as the heat travels from the surface to warm the cooler centre but, as the dough temperature reaches 43 °C, yeast activity begins to decrease and eventually ceases by 55 °C.

The dough piece surface dries and forms a relatively hard crust which tries to contain the expansion of the still relatively fluid dough centre. If sufficient gas is retained during these early phases of baking, then the pressure of the centre expansion can be sufficient to 'crack' the outer crust. This is the phenomenon that bakers call 'oven spring' and the degree of expansion is linked with both gas production and dough gas retention. Balancing the oven spring to give the desired degree of break requires careful integration of the formula and process factors which control it. In some products, e.g. sandwich bread, minimal oven spring is required while in others, e.g. baguettes, it is encouraged. However, in all cases it should be controlled.

Not all the air bubbles which are trapped within dough pieces need to survive the baking process. In some cases the trapped air may be squeezed out by pressures set up between the expanding dough and the walls of the pan holding it. Coalescence of gas bubbles can also occur during baking and small bubbles may merge to become larger bubbles which remain in the final product. If the dough rheology is unsuitable or if large volumes of carbon dioxide gas expansion are generated early in the baking process, then the crumb may be 'torn' apart. The resultant holes will be characterised by broken strands on crumb and rough sides.

Significant changes to the flour components also take place in the oven. Gelatinisation of the starch starts at about 60 °C and initially the starch granules absorb any free water in the dough. *Alpha*-amylase activity converts the starch into dextrins and then sugars and reaches its maximum activity between 60 °C and 70 °C. Too little amylase activity restricts loaf volume, because the starch structure becomes rigid too soon, while too much may cause the dough structure to become so fluid that the loaf collapses completely. The gluten network also begins to lose its elasticity and is no longer able to stretch to maintain dough expansion.

Some types of bread are characterised by the crispness of their crust, e.g. oven-bottom bread, hearth bread and baguettes. The first few moments in the oven are vital for the formation of a glossy crust. To obtain gloss, it is essential that vapour condenses on the surface to form a starch paste that will gelatinise, form dextrins and eventually caramelise to give both colour and shine. If there is excess water, *paste-type gelation* takes place while, with insufficient water, *crumb-type gelation* occurs. To deliver the necessary water, steam is introduced into the oven. Crusty-type breads made by the CBP require the introduction water or steam to achieve the desired effect.

Cooling and storage

After baking, many bread products are cooled and may be wrapped before consumption. The truly 'fresh' character of bread is lost within a few hours after leaving the oven as a number of changes take place in product character. Collectively such changes are referred to as staling and are related to changes in moisture content and distribution (Cauvain and Young, 2000) and the crystalline nature (retrogradation) of the starch (Pateras, 1998).

The progressive firming of breadcrumb with storage time is the one most commonly associated with staling of bread. The rate at which bread stales is independent of the breadmaking process used and linked with the moisture content of the product. Starch retrogradation proceeds more slowly in breads with higher crumb moisture contents. In products which are sold by fixed weight, this means that an increase in the moisture-to-solids ratio in the crumb, as well as giving an initially softer eating crumb, will stale at a lower rate. In the CBP the ability to increase the water-to-solids ratio in the dough while scaling at a fixed weight results in a loaf which stales more slowly and thus appears softer to the consumer over a longer period of time.

7

Optimising quality with the Chorleywood Bread Process

Primary causes of quality defects and their contributing factors

With all breadmaking processes and production there will be times when the final product quality does not reach the required level or when refinements and improvements are required. In order to optimise final product quality or to innovate and develop new bread products, it is necessary to have a thorough understanding of the influence of raw materials, the formulations used and the interaction with the processing conditions and equipment which might be employed. Making bread by the CBP presents no exception to such rules and, while it may have fewer processing steps than some breadmaking processes (e.g. it lacks the dough bulk fermentation process), the complex interactions (the rules) which characterise breadmaking must still be understood to ensure that the product quality is 'right'.

A systematic approach is required in order to optimise final product quality effectively and efficiently. Cauvain and Young (2001) provided a basic problem-solving guide and emphasised the importance of good records and a methodical approach to analysing the nature of the quality defect and how to set about remedial actions. They also identified a number of key ingredients and process factors which would affect final product quality, some of which are directly applicable to fermented products manufactured by the CBP.

Cauvain and Young (2001) described their problem-solving approach schematically as:

problem → primary cause → contributing factors → corrective action

Or, in more simple terms, as:

what is seen → why → because of ... → corrective action

This relationship between 'primary cause' and 'contributing factors' is very important in the optimisation of product quality and its use can reduce the time taken to identify remedial or corrective actions. The same approach allows for simplification of the way in which information for problem solving can be stored and presented. This aspect of quality optimisation is discussed further in Chapter 8.

There are a number of primary causes of quality defects that may occur in any bread or fermented product, whatever the process used to manufacture them, and a significant number of these occur for a variety of reasons. There are many primary causes of quality defects in the manufacture of fermented products but there are four which each have a significant number of contributing factors and provide a useful classification for simplifying the quality optimisation process. They are as follows:

1   Excess gas production.
2   Insufficient gas production.
3   Excessive gas retention in the dough.
4   Insufficient gas retention in the dough.

Contributing factors in the first two categories are all those which are commonly associated with the generation of carbon dioxide gas through bakers' yeast fermentation. In the manufacture of bread by the CBP these would be the following:

1   Yeast level: too much or too little (equating to excess and insufficient gas production respectively).
2   Dough temperature: too high or too low.
3   Intermediate proof time: too long or too short.
4   Proof temperature: too high or too low.
5   Final proof time: too long or too short.

The availability of substrate for the yeast can also contribute to gas production problems although in this case it would only do so through a lack of substrate, leading to insufficient gas production. This is not a very common effect given that there are already low levels of sugars in flour and these are commonly supplemented in breadmaking by the action of the amylases on damaged starch.

The production of gas is only one of the critical factors in breadmaking; the retention of the gas during the various fermentation stages and the early stages of baking is the other. As has been discussed previously and in many other publications (see for example Cauvain and Young (1998)), gas retention in dough depends on many individual and interactive factors. In the context of the CBP these may form 'subgroups' of contributing factors; for example we could list contributing factors as follows:

1   Flour protein content: too high or too low (equating to excess and insufficient gas retention/respectively).
2   Flour colour grade/ash/bran-scan level: too low or too high (the presence of bran has a negative factor on bread volume so that excessive gas retention comes from low-bran flours).
3   Improver level: too high or too low.
     (a) Fungal *alpha*-amylase level: too high or too low.
4   Incorrect type of improver. Here the reference would be to the correct level of improver being used but the composition of that improver is such that the formulation contributes too many or too high a proportion of functional ingredients, or too few or low a proportion of functional ingredients which give excessive or insufficient gas retention respectively (e.g. fungal *alpha*-amylase).

The few examples given above relating to dough gas retention illustrate the difficulties faced when using composite improvers of unknown formulation in the CBP (or any other breadmaking process for that matter). The problems are further compounded by the fact that not all ingredient and process contributions are simply a case of too much or too little. For example, as discussed in earlier sections the oxidising effect of ascorbic acid is limited by the availability of oxygen in the dough. Classically with the CBP it was considered and largely substantiated that over-oxidation with ascorbic acid was not a phenomenon that would be encountered. This was not the case when potassium bromate was used as an oxidant. The introduction of the pressure–vacuum mixer changed the potential relationship between available oxygen and the level of ascorbic acid such that it became possible to achieve a degree (albeit limited) of over-oxidation which is most commonly manifested in the final product as concavity of sidewalls or the top of lidded breads.

Quality defects in fermented products and their causes

In the discussion of quality defects in products manufactured by the CBP which follows, the approach has been to consider quality attributes in the order of shape, surface appearance, internal character, crumb texture, eating character, flavour and aroma. This is an arbitrary approach but follows what tends to happen in the bakery. Assuming that the dough has the appropriate character and is not rejected before processing commences, the first sign of a problem is when the product appearance changes. Inspection of internal features must await the passage of time until the bread is cool enough to be cut. Unfortunately an acceptable external appearance does not automatically mean an acceptable internal appearance. In some cases the reverse may be true but it is unlikely that a product with an unacceptable external appearance would be used even if the internal appearance was acceptable.

The major contributing factors to quality defects come from four main sources: ingredient qualities, ingredient quantities (the recipe), the processing conditions and interactions between the first three. The interactions in particular make the presentation of information related to quality optimisation/fault diagnosis complex and so in the tables which follow have been simplified on the basis of the influence of the following:

1   Ingredient qualities
2   Ingredient quantities
3   Mixing
4   Dough processing from dividing to final moulding
5   Proving and baking
6   Cooling, slicing (if used) and packing

Interactions have been indicated wherever possible.

Table 7.1 Quality defects associated with the quality of ingredients used in the CBP

Defect	Ingredient quality
Collapsed product (see Fig. 7.1)	Flour too weak, e.g. low protein or high ash/colour grade/bran-scan level, or both
	Improver too 'weak' (i.e. yielding poor dough gas retention)
	Combination of flour and improver too 'weak'
	Melting point of added fat too low
Excessive volume	Flour too strong, e.g. high protein content or low ash/colour grade/bran-scan level, or both
	Flour falling number too low
	Added enzyme activity too high
	Improver too 'strong' (i.e. yielding excessive dough gas retention)
	Combination of flour and improver too 'strong'
Lack of volume	Flour too weak, e.g. low protein or high ash/colour grade/bran-scan level, or both
	Flour falling number too high
	Enzyme activity too low
	Improver too 'weak' (i.e. yielding poor dough gas retention)
	Combination of flour and improver too 'weak'
	Melting point of added fat too low
	Yeast activity too low, e.g. poorly stored or out of date
Concave sides (keyholing)	Improver too 'strong' (i.e. yielding excessive dough gas retention)
	Combination of flour and improver too 'strong'
Concave top (lidded bread)	Improver too 'strong' (i.e. yielding excessive dough gas retention)
	Combination of flour and improver too 'strong'
Concave bottom in pan breads	Combination of flour and improver too 'strong'
Low shoulders (see Fig. 7.2)	Flour falling number too high
	Enzyme activity too low
	Improver too 'weak' (i.e. yielding poor dough gas retention)
	Combination of flour and improver too 'weak'
	Melting point of added fat too low
	Yeast activity too low, e.g. poorly stored or out of date
Sloping top	Melting point of added fat too low
Excessive oven spring	Improver too 'strong' (i.e. yielding excessive dough gas retention)
	Combination of flour and improver too 'strong'
Lack of oven spring	Improver too 'weak' (i.e. yielding poor dough gas retention)
	Combination of flour and improver too 'weak'
	Melting point of added fat too low
	Yeast activity too low, e.g. poorly stored or out of date
Blisters on top crust (see Fig. 7.3)	Improver too 'weak' (i.e. yielding poor dough gas retention)
	Combination of flour and improver too 'weak'
	Melting point of added fat too low
Cell structure too open	Improver too 'weak' (i.e. yielding poor dough gas retention)
	Combination of flour and improver too 'weak'
	Melting point of added fat too low
Cell structure uneven	Added enzyme activity too high
Bone (heavily compressed crumb)	Improver too 'strong' (i.e. yielding excessive dough gas retention)
	Combination of flour and improver too 'strong'
Large hole under crust	Melting point of added fat too low

Table 7.1 *Continued*

Defect	Ingredient quality
Hard patches (knots) in the crumb	Melting point of added fat too low
Gummy deposits on slicer blades	Flour falling number too low Added enzyme activity too high
Crumb texture too soft	Added enzyme activity too high
Crumb texture too firm	Added enzyme activity too high
Off aroma or flavour	Old fat or emulsifier

(a)   (b)

Fig. 7.1 (a) Collapsed product compared with (b) standard roll product (right).

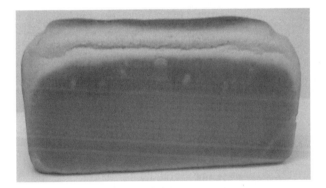

Fig. 7.2 Low shoulders with pan bread.

Fig. 7.3 Blister on bread.

Table 7.2   Quality defects associated with the CBP recipe (ingredient quantity)

Defect	Ingredient quantity
Collapsed product	Added fat level too low
	Improver level too low
Excessive volume	Level of *alpha*-amylase too high
	Improver level too high
	Yeast level too high (i.e. excess gas production)
	Salt level too low for yeast level
Lack of volume	Level of *alpha*-amylase too low
	Added fat level too low
	Yeast level too low (i.e. insufficient gas production)
	Improver level too low
	Sugar level too high
	Salt level too high for yeast level
	Dough water level too low
Concave side (keyholing)	Level of *alpha*-amylase too high
	Improver level too high
	Yeast level too high (i.e. excess gas production)
Concave top (lidded bread)	Level of *alpha*-amylase too high
	Improver level too high
	Yeast level too high (i.e. excess gas production)
Concave bottom in pan breads	Improver level too high
Low shoulders	Level of *alpha*-amylase too low
	Added fat level too low
	Yeast level too low (i.e. insufficient gas production)
	Improver level too low
	Dough water level too low
Excessive oven spring	Level of *alpha*-amylase too high
	Improver level too high
	Yeast level too high (i.e. excess gas production)
	Sugar level too low
	Salt level too low for yeast level
Lack of oven spring	Level of *alpha*-amylase too low
	Added fat level too low
	Yeast level too low (i.e. insufficient gas production)
	Improver level too low
	Sugar level too high
	Salt level too high for yeast level
	Dough water level too low
Flying top/wild break	Salt level too high for yeast level
	Dough water level too low
	Yeast level too low (i.e. insufficient gas production)
Blisters on top crust	Added fat level too low
	Improver level too low
	Yeast level too high (i.e. excess gas production)
Overhanging top (see Fig. 7.4)	Yeast level too high (i.e. excess gas production)
	Salt level too low for yeast level
	Dough water level too high
Sloping top	Level of added fat too low
	Improver level too low
Flat shape (oven-bottom and freestanding products)	Dough water level too high

Table 7.2 *Continued*

Defect	Ingredient quantity
Crust colour too dark	Sugar level too high
	Skimmed-milk powder level too high
Crust colour too pale	Sugar level too low
Cell structure too open	Level of added fat too low
	Improver level too low
	Yeast level too high
	Salt level too low for yeast level
Cell structure too close	Yeast level too low
	Salt level too high for yeast level
Cell structure uneven	Level of added fat too low
	Improver level too low
	Yeast level too high
Cell walls thick	Level of added fat too low
Crumb dull with streaks	Level of added fat too low
	Improver level too low
Bone (heavily compressed crumb)	Improver level too high
Large hole under crust	Level of added fat too low
	Improver level too low
	Yeast level too high
Gummy deposits on slicer blades	*Alpha*-amylase level too high
Crumb texture too soft	Level of added enzymes too high (e.g. maltogenic amylase)
	Yeast level too high
Crumb texture too firm	Level of added fat too low
	Improver level too low
	Salt level too high
Hard patches (knots) in the crumb	Level of added fat too low
	Improver level too low
Bread lacks flavour	Sugar level too low
	Salt level too low

Fig. 7.4 Overhanging top on bread.

Table 7.3 Quality defects associated with CBP mixing

Defect	Mixing conditions
Excessive volume	Energy input during mixing too high
	Mixer headspace pressure too high
	Final dough temperature too high
Lack of volume	Energy input during mixing too low
	Mixer headspace pressure too low (i.e. vacuum level too high)
	Final dough temperature too low
Concave sides (keyholing)	Energy input during mixing too high
	Mixer headspace pressure too high
Concave top (lidded bread)	Energy input during mixing too high
	Mixer headspace pressure too high
Low shoulders	Energy input during mixing too low
	Mixer headspace pressure too low (i.e. vacuum level too high)
	Final dough temperature too low
Sloping top	Energy input during mixing too low
	Mixer headspace pressure too low (i.e. vacuum level too high)
	Final dough temperature too low
Excessive oven spring	Mixer headspace pressure too high
Lack of oven spring	Energy input during mixing too low
	Mixer headspace pressure too low (i.e. vacuum level too high)
Blisters on top crust	Energy input during mixing too low
	Mixer headspace pressure too low (i.e. vacuum level too high)
	Final dough temperature too low
Overhanging top	Final dough temperature too high
Flying top	Final dough temperature too low
	Dough water level too low
Cell structure too open	Energy input during mixing too low
	Mixer headspace pressure too low (i.e. vacuum level too high)
	Final dough temperature too high
Cell structure uneven	Final dough temperature too high
Cell walls thick	Energy input during mixing too low
	Mixer headspace pressure too low (i.e. vacuum level too high)
	Final dough temperature too low
Streaks in the crumb	Energy input during mixing too low
	Final dough temperature too low
Bone (heavily compressed crumb)	Energy input during mixing too high
	Mixer headspace pressure too high
Large hole under crust	Energy input during mixing too low
	Mixer headspace pressure too low (i.e. vacuum level too high)
	Final dough temperature too low
	Final dough temperature too high
Crumb texture too firm	Energy input during mixing too low
	Mixer headspace pressure too low (i.e. vacuum level too high)
	Final dough temperature too low
Off aroma or flavour	Addition of old dough at the mixer

Table 7.4 Quality defects associated with CBP dough processing: 1

Defect	Dividing	First moulding
Excessive volume	Divider weight settings too high	
	Dough piece too large for pan	
Lack of volume	Divider weight settings too low	
	Dough piece too small for pan	
Concave sides (keyholing)	Divider weight settings too high	
Concave top (lidded bread)	Divider weight settings too low	
Low shoulders	Divider weight settings too low	
Overhanging top	Divider weight settings too high	
	Dough piece too large for pan	
Cell structure uneven		Excessive pressure during first moulding
Streaks in the crumb		Excessive pressure during first moulding
Small random holes	Excess divider oil	Skinning on the first moulder

Table 7.5   Quality defects associated with CBP dough processing: 2

Defect	First proof	Final moulding
Lack of volume	First proof time too short	
Sloping top		Poor placing of dough pieces in pans
		Uneven cylinder formed during final moulding
Low shoulders		Uneven four-piecing after final moulding
Blisters/bubbles on top crust		Damage during final moulding
White patches or scorch marks on the sides of pan breads		Pans too hot when dough pieces placed in them
Cell structure too open	First proof time too long	
Cell structure too close	First proof time too short	
Cell structure uneven		Damage during final moulding
Streaks in the crumb		Damage during final moulding
Small random holes	First proof humidity too low (skinning)	
Large hole under crust		Damage during final moulding (pressure board)
Holes along the moulding line		Poor curling during final moulding

Table 7.6 Quality defects associated with proving and baking CBP doughs

Defect	Proving	Baking
Collapsed product		Bumpy transfer from prover to oven Excess steaming
Excessive volume	Proof time too long Proof temperature too high	Baking temperature too low
Lack of volume	Proof time too short Proof humidity too low (skinning) Proof temperature too low	Baking temperature too high Insufficient steaming
Flat shape to oven-bottom and freestanding products	Proof humidity too high (flow)	Excess steaming
Concave bottom in pan breads		Too much bottom heat during baking
Low shoulders	Proof time too short Proof humidity too high (flow)	Proof temperature too high
Excessive oven spring		Baking temperature too low
Lack of oven spring		Insufficient steaming
Blisters on top crust	Proof time too long Condensation in the final prover	
Overhanging top with pan breads	Proof time too long Proof humidity too high Proof temperature too high	Baking temperature too low
Flying top	Proof time too short Proof humidity too high (flow)	Baking temperature too high Insufficient steaming
Crust colour too dark		Baking temperature too high Baking time too long
Crust colour too pale	Proof humidity too low (skinning)	Baking temperature too low Baking time too short
Crust too thick	Proof humidity too low (skinning)	Baking time too long
Cell structure too open	Proof time too long Proof temperature too high	Baking temperature too low
Cell structure too close	Proof time too short Proof humidity too high (flow)	
Cell structure uneven	Proof time too long	
Bone (heavily compressed crumb)		Baking temperature too low

Table 7.6 *Continued*

Defect	Proving	Baking
Large hole under crust	Proof humidity too low (skinning)	
Large hole toward the base		Too much bottom heat (pan breads)
Gummy deposits on slicer blades		Baking temperature–time combination incorrect
Crumb texture too soft		Baking time too short
Crumb texture too firm		Baking time too long
Burned aroma		Baking temperature too high

Table 7.7 Quality defects associated with cooling and storing bread made by the CBP

Defect	Cooling	Storage
Concave sides (keyholing)	Cooling time too long	
Concave top (lidded bread)	Cooling time too long	
Gummy deposits on slicer blades	Cooling time too short	
Crumb texture too firm	Cooling time too long	Storage time too long Storage temperature too low
Tough leathery crust	Cooling time too short	
Product goes mouldy quickly	Product wrapped too warm (condensation)	Storage temperature too high (condensation)
Product has sweet sickly pineapple smell (rope)	Product wrapped too warm (condensation)	Storage temperature too high (condensation)

Quality defects in four-piece bread

The manufacture of four-piece bread has been discussed in Chapter 6. This moulding and panning technique is commonly used for the majority of lidded sandwich bread manufactured in the UK and elsewhere. It is most effectively produced using the CBP and is so ubiquitous that lidded sandwich bread has come to be seen as the only product that can be produced by the CBP. This is far from the case as has already been discussed and is fully illustrated in Chapter 9.

The reorientation of the four small pieces cut from one long cylinder can make the diagnosis of quality defects more complicated than would be seen with single-piece dough moulding. The potential for damage to bread cell structure has been discussed in Chapter 6. Commonly cell structure damage shows itself as areas or streaks of darker crumb in which the cell walls appear thicker and the average cell size is larger. Bread manufactured using the four-piece moulding technique may show such streaks, swirls and dark patches, often seemingly random in distribution along the length of the loaf or from slice to slice.

In fact, bread manufactured using four-piece moulding if the pieces are optimally located in the pan exhibits a systematic structure along the length of the loaf. Essentially in the centre of each of the four pieces the orientation of gas cells in the dough and subsequent bread is horizontal. This arises because of the action of the sheeting rolls in the final moulder. After being placed in the pan, the natural movement of gas bubbles is to expand in all directions fairly equally until the pan is filled and then the natural movement is upwards. In contrast, at the joins of the four pieces, dough expansion quickly moves the pieces into contact with one another and the pressure for upward expansion occurs earlier during the proving cycle. The end result of the four-piece technique and the pressures on the gas bubbles is to create alternating horizontal (centre of a piece) with vertical (join of two pieces) cell structures. In the areas of crumb between the two main cell structure types the cell structure orientation gradually changes from horizontal to vertical and back again along the length of the pan. A knowledge of this systematic variation in cell structure is invaluable in identifying the source of quality defects in four-piece bread.

8

Knowledge-based systems and the Chorleywood Bread Process

Introduction

Since the advent of the CBP, computing technology has moved at a rapid pace. Computing science has been used in all aspects of baking, many of which have helped to make the technology of breadmaking and the businesses in which it is undertaken more efficient and easier to operate and manage. An area such as database technology has enabled bakers to maintain a record of their recipes, their ingredient stocks and processing records. Spreadsheets have enabled the bakery to manage its finances. Word processing has allowed the baker to keep up with the many legislative practices, health and safety, traceability and due diligence required when running a bakery. Other computing technologies, namely image analysis (Whitworth *et al.*, 2005), X-ray tomography (Whitworth and Alava, 1999), magnetic resonance imaging (Tiphaine *et al.*, 2005) and data mining (Wilson *et al.*, 2005), have helped in the understanding of the basic science and the interaction of machine with bread product. Process control harnessing computing technology has revolutionised the plant bakery and has enabled products to be produced uniformly day after day and computers with their integrated control systems often control the daily production of bread with very little intervention from the plant operative.

Knowledge-based systems are another such field of information technology which has helped the bakery technologist. Since its introduction in the 1980s, its application to baking technology has been successful. Knowledge-based systems encompass knowledge about a domain or subject area and the knowledge is encoded in a computer program so that it might be used by others, perhaps the less experienced, to obtain answers to problems in that domain. The answers might be connected with changes in processing or product formulation; experimenting at the personal computer (PC) with the question 'what if I . . . ?' can yield solutions without the expense or time involved in trying out the change in the test bakery or on the plant. Often the solutions are supported by the 'why this happens' information, thus expanding the user's own knowledge in a passive way.

Advances in knowledge and know-how about breadmaking, whether that be by the traditional methods or the CBP, have likewise moved forwards and the understanding of the science and technology of the products and the process by which they are made are better understood by technologists, bakers and scientists. Knowledge about breadmaking has evolved over time. Since breadmaking began, knowledge about the products and their process requirements was passed down by word of mouth and actions, later by the written word and illustrations and latterly by moving images and commentary on film, CD ROM and DVD.

The next generation of bakers has a hard task to absorb both the theory and the practice of breadmaking. There is no longer the luxury of observing the master baker at work or the time to assimilate all the changes that occur in the rheology of the dough piece as it travels through the processing to become the finished product. In the UK and other parts of the world, bakery colleges are disappearing as students no longer choose the more manual trades on which to base their careers. Computer tools in the form of knowledge-based systems can go some way to filling the knowledge and training gaps which exist in a format much loved by today's newcomers to the world of baking technology.

Knowledge-based systems can only be developed if knowledge about the technology of the product is known and made available. Knowledge about bread technology, and in particular that for the CBP, can be represented in a knowledge-based system. The way that it is represented and structured takes many forms. Some of these will be described shortly, but first it is prudent to offer some definitions about knowledge-based systems and 'expert' systems as they were first called on the introduction of this branch of computing technology.

An expert system is a computer program which attempts to model the expertise of a human expert within a specific domain.

A knowledge-based system incorporates heuristic knowledge (rules of thumb, best guess, intuitive modelling, directional changes, etc.) captured from intelligent sources of knowledge.

For the purposes of this chapter the term 'knowledge-based systems' will be used to include both definitions as often the terms are used interchangeably. In the pioneering days of this computing technology the term 'expert system' was overused and claims about its value to the computing world were difficult to realise in the climate of those early days (1980s) when the integration of words, images, equations, etc., was not as easy as it is in today's Microsoft Windows™ or Macintosh environments. However, even in their early form, provided that the domain area was clearly specified, they proved useful tools to assist the baker in perfecting a faulty or below-standard product.

To illustrate the difference between a data-processing program and a knowledge-based system, consider the following example. Suppose that a baker has a product recipe with ingredients and their quantities and wants to develop the nutritional breakdown of the finished product for the product label. If the nutritional composition, e.g. carbohydrates, fats and proteins, of each ingredient is known, then by multiplying the ingredient quantity by its nutritional component, e.g. fat for each ingredient, and then summing the contribution of fat say from each individual ingredient, the total fat for the product can be found. The path followed through a data-base program is predictable for each nutritional component. If the baker follows a recipe and processing method, say for French sticks, which results in a product that is faulty or below standard, then in order to obtain some advice for corrective actions the computer program that the baker uses must contain rules and facts about the ingredients, the processing and especially the interactions between them. When a problem is encountered with another product, say pan bread, then the sets of rules and path through the program will differ because ingredient, recipe and process interactions tend to be product specific. The route through the program is dependent upon the information given, the decisions made by the user and the knowledge contained and defined within the system. Explanation of the causes of the faulty problem and corrective actions can be obtained by the user.

The golden rules for developing such systems can be summarised as follows:

1   The domain knowledge must exist and be available.
2   The integrity and validity of the knowledge must be sound.
3   The domain to be tackled should be of manageable size or capable of being divided into manageable portions. These portions can be linked to extend the domain if necessary.
4   The structuring of the knowledge is very important.
5   Design of the screen interface should be easy to use and uncluttered.
6   Because of the unpredictable routes through the program, testing should be extensive and multifaceted.
7   Documentation and user guide (help file) should be comprehensive, extensive and easily accessible.

Knowledge about the chosen domain can be from many sources, human expert, literature, records, etc. Once the knowledge has been gathered and encoded into a knowledge-based system, then it can be used by technologists with different skill levels. No matter who uses the system, expert or novice, the system never tires or forgets and the advice and answers provided to the user are consistent. Using the information given, users can expand their own knowledge of baking technology.

The Bread Faults Expert System

In 1988 the UK Ministry of Agriculture, Fisheries and Food (MAFF) (now Department for Environment, Food and Rural Affairs (DEFRA)) was keen to introduce new computing technologies, and in particular 'expert systems' to the food industry. Scientists at the FMBRA, Chorleywood, were not slow to realise the potential that such computing technology might offer to the baking industry. Whilst the CBP was by then almost 30 years old, the understanding of some of the technology, particularly when things went wrong with the product made by the process, was lacking. A fault diagnosis system for breads made by the CBP could go some way to redressing this lack of understanding and could augment the technologists' understanding of the process. Thus funding was secured to develop a knowledge-based system for the diagnosis of faults in 800 g white breads made by the CBP (Young, 1991). The system was developed at the FMBRA. The domain of a single size and type of loaf may seem very narrow by today's standards but was considered an achievable goal for a new, and untried in a bakery context, computing technology. At the end of the project the FMBRA made the system commercially available to the baking industry. In fact, this system was one of the first such systems to be developed for the baking industry worldwide.

Breads made by the CBP was the domain area chosen for the study. It was developed over a period of 18 months in 1988–90. Aspects of the accumulated knowledge of the CBP both in the literature and held by experts was made available. The experts in the CBP involved were Bill Collins, Stanley Cauvain and other technologists at the FMBRA. The systems designer and programmer was Linda Young. The baking trials and photography were undertaken by Kim Little.

The research objective was to determine whether the knowledge-based systems area of computing science could be applied for the benefit of the baking industry. At the end of the project the potential for such systems was soon realised and the objective was extended to produce a commercially available computer program for the baking industry. Knowledge for the system was elicited by Linda Young and Tom Fearn from Bill Collins and Stanley Cauvain by means of structured interviews and recorded on magnetic tape. Supporting background information was acquired from books and reports available in the FMBRA library. Knowledge of the faults was underpinned by appropriate baking trials and a photographic record of the manifestations of the faults was made. Initially a survey was undertaken of the most common faults in CBP bread production. However, it soon became obvious that including all bread varieties made by the CBP was too ambitious and the decision was taken to limit the system to faults in white pan breads (at that time, 50% of the bread market). This was considered a manageable in-depth domain and would provide a useful diagnostic tool for the industry.

Because of the cost of computing hardware and the limited availability of mass computer storage at that time, the option of including images in the software was dismissed as being too expensive and time consuming for the project. Instead a handbook was produced to go with the software and this contained descriptors and images of the faults included in the system. The completed system marketed as 'the Bread Faults Expert System' sold for £900.

The success of the Bread Faults Expert System and the use of knowledge-based systems to underpin and provide knowledge about baking technology stimulated development of other systems for the baking industry. The funding provided by MAFF had served its original purpose, namely to demonstrate that the knowledge-based systems area of computing science could be applied for the benefit of the baking industry. Systems were subsequently developed for the flour confectionery domain (a cake expert system with modules ERH CALC (Young and Cauvain, 1994), BALANCE (Young et al., 1998) and Fault DoC (Petryszak et al., 1995), for retarding, for hot cross bun production, etc.). Funding for these systems came from the FMBRA research programme and was funded by the members of FMBRA.

Intelligent Mediator and Bread Advisor

Ten years later and as computing technology advanced into the Internet age, MAFF again funded work to show the potential to the baking industry of using knowledge systems delivered via this medium. The Intelligent Mediator was developed as a demonstrator program for enabling knowledge about breadmaking to be available interactively via the Internet. Intelligent Mediator had Bread Advisor (Young and Cauvain, 2001) as a 'back-end' knowledge-based program available for use over the Internet. This knowledge-based system encompassed far more than the original Bread Faults Expert System and focused now on breads made by the five major breadmaking processes, one of which was the CBP; 14 generic types of bread product were included. Consequently the Bread Advisor is a much more comprehensive knowledge system than the original Bread Faults Expert System and includes not only fault diagnosis but also aspects about processing and experimentation at the PC by means of a 'what if?' module.

At this point it is appropriate to describe the functionality of the Bread Advisor (Fig. 8.1) as it pertains to the CBP.

Choosing a product and process
The system can be consulted and used by choosing one product type made by a processing method. During the consultation, these two choices define a product profile which is built and carried forwards to other parts of the

Fig. 8.1   Start screen.

system. The product profile holds the information needed by each of the elements of the system as they are reached. At start-up the user can select the item Chorleywood Bread Process from a window (Fig. 8.2) containing the main breadmaking processes, including those such as No-time Dough, Bulk Fermentation Process, Sponge and Dough Process or Flour Brew

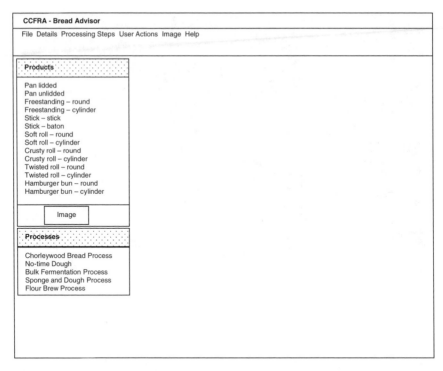

Fig. 8.2   Products and processes.

Process. Generic product types include pan breads, lidded or unlidded, free-standing (e.g. oven-bottom or hearth) breads, sticks (e.g. baguettes), soft rolls, crusty rolls, twisted rolls and hamburger buns. The generic types are representative of the major product types produced throughout the world and in effect represent products of different dimension types. Images can be viewed of the product type highlighted (Fig. 8.3).

Once the product is chosen with CBP, the consultation can take the form of one of the following:

1 Fault diagnosis or quality enhancement.
2 'What if?' experimentation.
3 Examination of process details.

Fault diagnosis or quality enhancement
Choosing the icon for fault diagnosis, a screen is displayed showing the categories of faults (Fig. 8.4) and the actual faults that might occur for the category in question (Fig. 8.5).

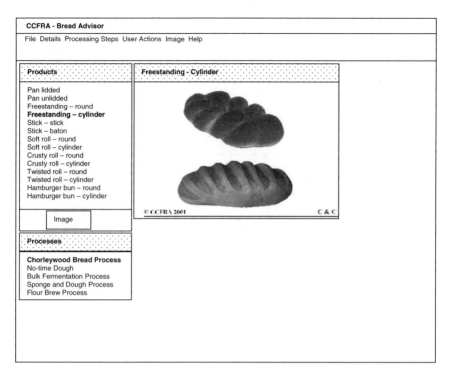

Fig. 8.3 Example product image.

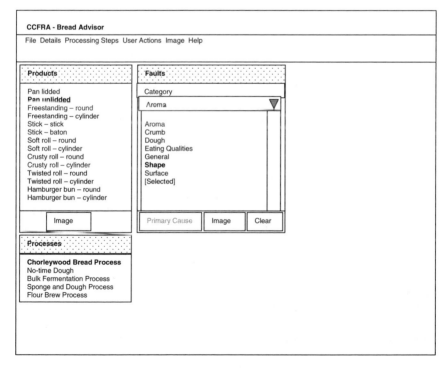

Fig. 8.4 Fault categories.

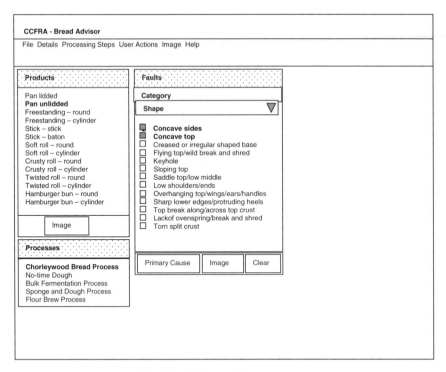

Fig. 8.5   Faults within a category.

The categories are diverse and include aroma, crumb (faults which occur inside the product, e.g. holes, texture, structure and colour), dough (faults which occur during the processing of the dough, e.g. sticky or soft dough), eating qualities (for both crumb and crust), flavour, shape (e.g. concavity, low shoulders and lack of oven spring) and surface (e.g. crust colour, spots, blisters and wrinkles). Faults such as low volume and collapsed product are included in a 'catch-all' category called General. Dividing the categories in this way enables the user to find quickly the fault in question. If the product is exhibiting several faults, then they can each be selected for a diagnosis. As an *aide-mémoire* the selected faults can be seen if the Selected category is shown (Fig. 8.6). The software is intended for use internationally. As the naming of faults is often unique to the country and product, an image of a fault can be seen by clicking the fault in question and then viewing it by clicking the image button (Fig. 8.7).

Faults or product deficiencies rarely have a single cause. However, they can be split into those which are considered 'primary' and those which have 'contributed' to the faults in question. For this reason a user can easily and quickly obtain them by clicking the Primary Causes button. This action

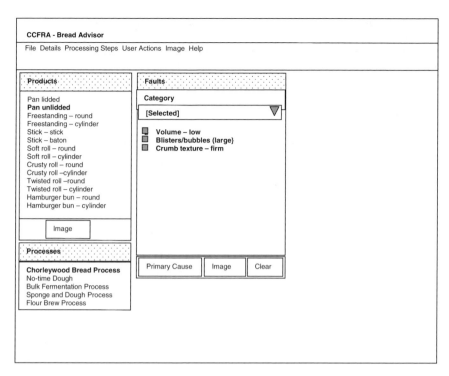

Fig. 8.6 Faults-selected category.

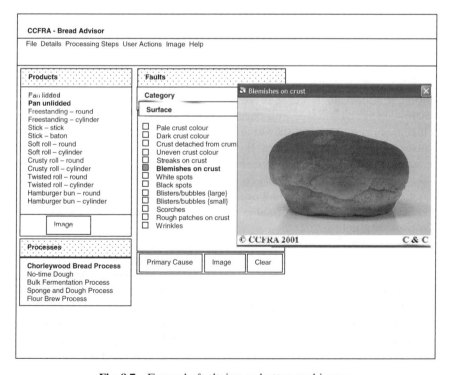

Fig. 8.7 Example fault: image button and image.

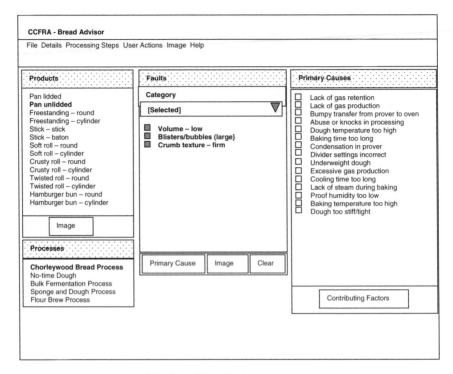

Fig. 8.8　List of primary causes.

reveals a further pop-up window with such causes ranked in order of likelihood, the most likely being listed at the top (Fig. 8.8). The list of primary causes can be considered and checked out with the processing conditions which occurred. Any factors which might have contributed to any of these causes can be displayed when the cause itself is checked and the Contributing Factors button selected (Fig. 8.9). At the end of a consultation the baker has a 'suspects' list which can be investigated together with the local circumstances that the product underwent to reveal the cause or causes of the fault. As a result the necessary corrective action can be taken to improve the product quality.

For an experienced baker the fault diagnostic aspect of the Bread Advisor considers all the necessary information and offers a quick and thorough investigation of the possible causes known to produce the faults in breadmaking. Unlike a human, the software never forgets or overlooks a possible cause. For novice bakers the same aspect offers knowledge about the causes of faults from which they can build their own knowledge base about bread faults. Suspects can be eliminated more quickly when processing conditions are checked and the 'once in a lifetime' fault flagged for investigation.

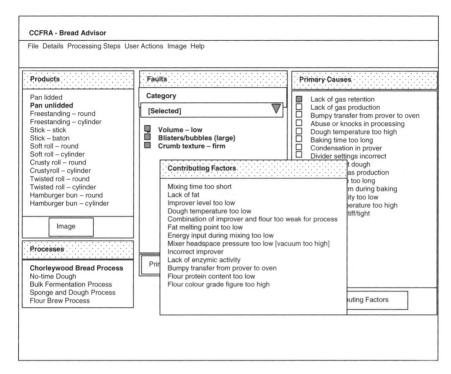

Fig. 8.9   List of contributing factors.

'What if?' experimentation

It is often useful to pose the question 'what if ... ?' when learning about the breadmaking process or attempting to change the quality of a product, perhaps to improve it or to explore a new product concept. The 'what if?' functionality allows the user to experiment with the process settings and to get feedback on the results. After selecting the What if? icon, the user can select a process step for investigation. One of the process steps which differentiates the CBP from others is mixing. Here and in any of the processing steps chosen a list of 'what if' questions is displayed (Fig. 8.10), e.g. What if I increase the mixer headspace pressure? By selecting the Query button, the resulting change to product quality will be given (Fig. 8.11). For example, in the case of the question about mixer headspace, the answer will be 'The cell structure will become more open (larger average size)'. This aspect of the software helps the novice user to learn quickly the consequences of any actions which might be taken on the plant when using the CBP or indeed any other breadmaking process.

Processing details

In addition to the 'what-if?' experimentation, any of the processing steps for the product can be investigated and the parameters relevant to the

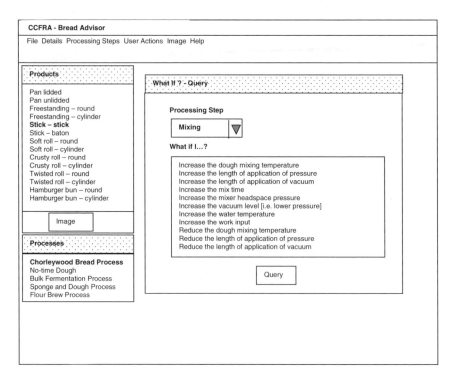

Fig. 8.10 'What if?' query: questions.

Fig. 8.11 'What if?' query: advice.

chosen step entered for comparison with generic settings for the step (Fig. 8.12). The generic settings are only those relevant to the process chosen. Settings for CBP can be checked out and, where they are at variance with the requirements for the process and product, a message is displayed giving the range of values in which they should lie to achieve acceptable product quality together with a message giving details of consequences to the product quality (Fig. 8.13). This facility is very useful to the novice who may be unsure of the settings required for the CBP.

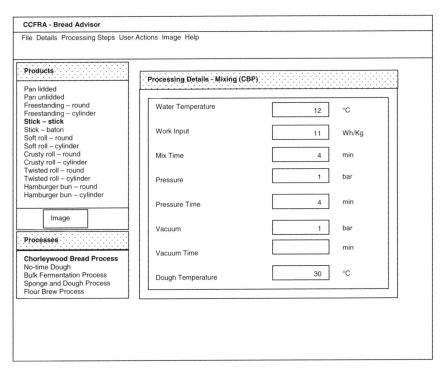

Fig. 8.12 Process step settings check.

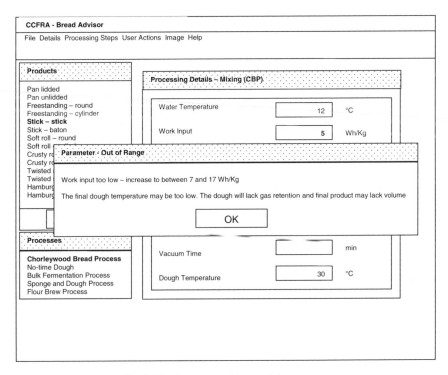

Fig. 8.13 Process settings advice screen.

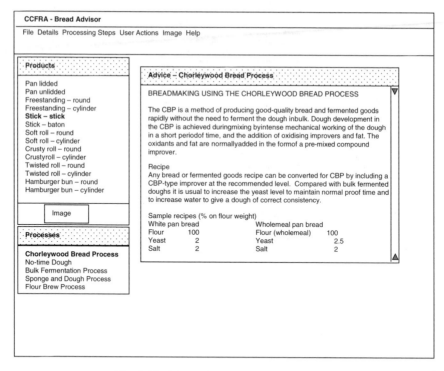

Fig. 8.14   Example process advice screen.

Advice and help

General advice about the CBP (Fig. 8.14) can be displayed and, with the assistance of the Bread Advisor Help file (Fig. 8.15) at any stage, users can learn how to navigate around the software and to obtain answers about the CBP processing steps.

Using knowledge-based systems as teaching tools

The Bread Advisor software can be used in a structured way in a classroom environment. For example, the tutor may set a series of questions relevant to the fermented products being studied. The student could harness the system to determine the answers and solutions. For example, what causes blisters or bubbles on a product surface? The student would select this fault for the bread type and process method, obtain the suspects list and investigate each in turn, looking at the guideline parameter ranges for any processing steps or ingredients indicated as possible causes. Finally the student could investigate directional changes which might be made to the ingredients or the processing steps to cure the fault and could try out the ideas

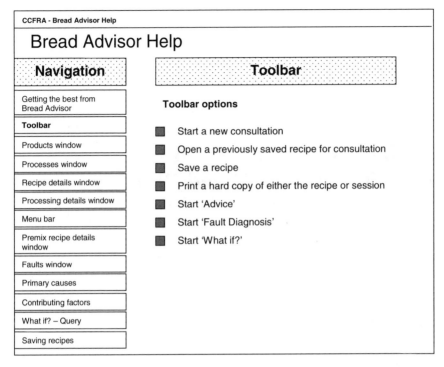

Fig. 8.15   Example help screen.

suggested in a training bakery, thus re-enforcing the theory with the practical. Using the medium of computing alongside the practical elements can consolidate and expand the student's knowledge of the fermented product.

Conclusions

The innovation leading to the invention of the CBP did not stop with the practical baking science aspects. The same wish to transfer the know-how into a wider commercial forum was manifested with the development of knowledge-based systems. In the first instance this was with a diagnostic tool for pan breads made by the CBP and pointed the way to the development of other knowledge systems for the baking industry. The development of the Bread Advisor system as both a 'back-end' program for use in a Web or Internet environment and as a stand-alone system has made knowledge about the CBP available in a modern medium and to a much wider audience. Use of such software alongside the practical implementation and operation of the process can empower today's bakers to develop new products whilst enlisting the versatility of the CBP.

9

Variety bread production with the Chorleywood Bread Process

Introduction

Most fermented products produced by traditional mixing methods or equipment can also be produced using the CBP and high-speed mixers capable of delivering the energy input on the timescales required for the process. The CBP does not rely on special recipes and any normal recipe can be used provided that some adjustments are made for the ingredients in order to take into account the essential features of the process. These recipe changes are relatively simple to implement.

In the development of the CBP, many combinations of the oxidising agents and other functional ingredients were evaluated before arriving at the optimum combinations. The first publication to summarise CBP recipes for the production of a variety of bread types was *BBIRA Report 74* (Collins, 1964). At that time, potassium bromate was a permitted oxidising agent in the manufacture of bread in the UK and most flours that were used in bakeries came from the flour mill with approximately 15 ppm potassium bromate present. Thus, the first published CBP recipes were based on a mixture of ascorbic acid and potassium bromate. Later the level of potassium bromate used in the CBP was increased following work which showed that there was a synergy between ascorbic acid and potassium bromate (Chamberlain and Collins, 1981) so that a 'typical' oxidising improver became based on 30 ppm potassium bromate and 45 ppm ascorbic acid. This total of 75 ppm 'oxidants' yielded similar results to the original 15 ppm potassium bromate and 75 ppm ascorbic acid.

As UK legislation changed to remove potassium bromate (and later azodicarbonamide) from the list of permitted bread improvers, typical UK improvers became based exclusively on ascorbic acid and the levels of other functional ingredients (e.g. enzymes) were adjusted to compensate for the lack of potassium bromate. Thus, in the recipes which are included in this section, improver formulations would be typically based on the use of only ascorbic acid but, in those parts of the world where potassium bromate and other oxidising agents are permitted, improver formulations will vary.

Changing recipes to make products using the Chorleywood Bread Process

1   When received at the bakery, white UK flour may contain ascorbic acid and enzymes, such as fungal *alpha*-amylase. In other parts of the world, other oxidants may be added and flours may be bleached, e.g. with benzoyl peroxide. Flour treatment should be checked before deciding on any recipe modifications. The addition of an oxidising agent, ascorbic acid (vitamin C) at a level of 75–200 ppm (depending on mixing conditions) or a CBP-type proprietary oxidising improver at the recommended level, is essential.

2   Originally a simple method for the addition of ascorbic acid was to make a solution of 5 oz (141.75 g) ascorbic acid to 15 pints (8.63 l) water and to use 1 pint (0.575 l) of this solution per sack (127 kg) of flour. With the advent of liquid improvers, ascorbic acid can now be included in solution form at a rate of up to 200 ppm (equivalent to 0.0002 kg of ascorbic acid in 1 kg of flour). It must be remembered that the water in the liquid improver should be deducted from the total amount of water added to the mix so that no errors are made in added water levels (see Chapter 4).

3   Fat or an emulsifying agent or a blend of the two should be included as an essential ingredient. The recommended level for fat is 0.7%. This fat should be one specially recommended for CBP with a suitable melting point profile (see Chapter 4).

It should be remembered that an emulsifying agent on its own will *not* provide the same functionality as fat although the addition of either or both will typically lead to an improvement in gas retention.

4   CBP-type improvers which contain both the oxidising agent and the fat are a very convenient way of dealing with these requirements. In such cases the rate of addition of the improver is typically between 1 and 2% of flour weight in the UK although lower levels, e.g. 0.5% flour weight, may be used in parts of the world.

5   Extra water should be included above what is used in a bulk fermentation recipe, equivalent to about (one extra gallon per sack) 3.62 l per 100 kg of flour. The extra water is necessary so that the dough rheology and consistency will be softer and will process satisfactorily through the plant. The extra water equates to the softening which occurs during bulk

fermentation. If a no-time spiral mixed recipe is being adapted for use with CBP, then the addition of extra water may be lower, typically 11 per 100 kg of flour.

6 In order to maintain proof times suitable for the available plant capacity, extra yeast is required. In UK practice this increase is 50–100% of the yeast level used in a bulk fermentation recipe. When adapting no-time spiral mixed recipes, then little or no adjustment of yeast level will be required.

Converting bulk fermentation recipes to the Chorleywood Bread Process

Special recipes are not required for products made by the CBP and existing recipes may be adapted to allow for the special benefits from the process.

The following example recipe shows the principles associated with conversion.

Table 9.1 Converting bulk fermentation recipes to the CBP

Ingredient	Amount					
	3h bulk fermentation			CBP		
	(%)	(kg)	(lb)	(%)	(kg)	(lb)
Flour	100	100	280	100	100	280
Yeast	1.1	1.1	3	2.1	2.1	6
Salt	2.1	2.1	6	2.1	2.1	6
Fat	0.7	0.7	2	0.7	0.7	2
Water	57.1	57.1	160	60.7	60.7	170
Soya flour (full fat, enzyme active)	0.7	0.7	2	0.7	0.7	2
Ascorbic acid	–	–	Nil	0.02	0.02	9.6 g

Steps to convert a traditional recipe to a Chorleywood Bread Process recipe

1 When converting recipes which were originally expressed in imperial units, it is easiest to convert the imperial recipe to 'bakers' per cent' and then to apply the changes for CBP. As 'bakers' per cent', the flour is expressed as 100% and other ingredients are specified as a percentage of the flour, i.e. each ingredient is expressed as a percentage of the flour weight used. For example, in Table 9.2, 280 lb of flour is converted to 100%. The 3 lb of yeast converts to $3/280 \times 100 = 1.1\%$ and so on.

Table 9.2 Steps to convert a traditional recipe to a CBP recipe

	Weight in imperial units (traditional recipe) (lb)	Bakers' per cent (%)	Bakers' per cent converted for CBP (%)	Weight in metric units* (kg)
Flour	280	100	100	127
Yeast	3	1.1	2.1	2.7
Salt	6	2.1	2.1	2.7
Fat	2	0.7	0.7	0.9
Water	160	57.1	60.7	77.1
Soya flour (full fat, enzyme active)	2	0.7	0.7	0.9
Ascorbic acid	0	–	0.02	0.03

* The values in this column have been rounded up to facilitate the weighing of ingredients in practical situations (e.g. quantity of fat = 0.7 × 280 / 100 = 1.96 kg (rounded to 2 kg)).

2   Using these bakers' per cent figures, the remaining ingredients can be adjusted for the CBP.
3   The other ingredients need to be adjusted as follows:
   (a)   Increase yeast by 50–100% to maintain proof times for plant capacity.
   (b)   Adjust fat to 0.7% of flour weight.
   (c)   Increase water by approximately 3.5%.
   (d)   Add oxidising improver, ascorbic acid, at a rate of 75 ppm.

A simple spreadsheet program could be written to convert traditional recipes using the rules explained above.

Useful conversion information
The list below gives conversions from imperial or traditional recipes to metric values used today:

Imperial	*Metric*
1 lb	453.592 37 g
1 lb	0.453 592 37 kg
1 oz	28.35 g
1 cwt	50.802 kg
1 pint	575 ml
1 gallon	4.6 l
1 cup	250 ml
1 tablespoon	15 ml
1 dessertspoon	10 ml
1 teaspoon	5 ml

Other statistics from older recipes are:

$$1 \text{ sack} = 280 \text{ lb} = 127 \text{ kg}$$

When converting recipes, care must be taken when calculating and rounding off for the small quantities of ingredients. In practical terms for the conversion from pounds to grams the figure 454 g (1 lb) is generally used. In the recipes given we have expressed the quantities in percentages based on flour weight. The per cent column can be used to produce the metric quantities. All recipes given are intended as a base point from which to apply local conditions, e.g. the adjustment of water dependent upon the flour batch and its water absorption characteristics.

The following scaling weights are approximate as the moisture loss during baking will depend on baking temperature and time, ratio of product surface area to volume, and oven type. The following values are intended as a guide:

920 g for 800 g pan bread
460 g for 400 g pan bread
550 g for 400 g pain Parisien
250 g for 210 g baton
350 g for 250 g French stick or baguette
2000 g 'head or press weight' for 36 × 56 g rolls (55 g baked)

Note that, with respect to the pain Parisien, in the UK bread should be sold as 400 g and multiples thereof (HMSO, 1988) and, with respect to batons, French sticks and baguettes, in the UK breads below 300 g are not subject to weight legislation.

Using improvers

The fat, soya flour, enzyme-active materials and ascorbic acid can be replaced with a proprietary improver. A general formulation for such an improver for a recipe using bakers' per cent is:

fat	0.7%
soya flour (full-fat enzyme active)	0.7%
enzyme-active materials, ascorbic acid	0.02%

Note that some improvers may not contain soya flour which may have been substituted with wheat flour or starch.

An improver containing the above ingredients can be added at a rate of 1.0–2.0% of flour weight for bread and morning goods respectively.

Where small quantities of other ingredients are required, e.g. skimmed-milk powder, malt flour or emulsifiers, these may be included for convenience in proprietary improvers. It is advantageous to know what ingredients are included in an improver so that no unexpected problems are manifested in the end product because of under-dosing or over-dosing of ingredients. It is recommended that bakers discuss their requirements with the improver supplier to ensure that the additive is suitable for the bread quality required and the plant used.

The quantity of water given in each of the recipes which follow is intended as a guide and because of the varying nature of the flour will vary a little (usually between 1% and 2% of the numbers given). This quantity will depend on the flour moisture content, damaged starch and water absorption values and are discussed fully in Chapter 4.

Use of rework

In many bakeries there is often a small amount of dough from a batch mix left over after scaling and rather than waste the dough it is reprocessed with the next batch mix. In the CBP this rework dough should be fed back into the mixing machine. However, this rework should be kept to a minimum and at most to 10% of flour weight for the new mix. The weight of this added dough should be excluded when calculating the energy requirement for the mix. The consequences of excess rework of dough in a batch are irregular cell structure and poor slice appearance.

Using one type of rework in another form of dough should be avoided. Clearly brown, wholemeal or mixed-grain bread doughs should not be reused with white doughs. Wholemeal bread should only contain wholemeal rework since the ingredients permitted in the manufacture of wholemeal bread are commonly limited by legislation in different countries.

Storing rework and using it at another time may cause quality defects, such as loss of volume and poor irregular cell structure. If stored for any significant length of time, e.g. a few hours, the high dough temperature may lead to the development of unwanted 'off-odours' or flavours in the final product. Aged rework should not be seen as a potential substitute for a sponge to introduce 'flavour' unless it has been held under controlled conditions.

The reworking of old bread should never be carried out. Such practices can lead to serious problems with microbial contamination, especially rope (Cauvain and Young, 2001).

Dough consistency

The importance of dough temperature and consistency is described in Chapter 5. In the absence of a watt–hour meter, for regular production of one type of dough, dough consistency can be controlled by recording the time taken to impart a constant work input to a constant dough weight at a constant dough temperature. Under these circumstances a decrease in the recorded time indicates a 'tighter' dough and suggests that the water absorption requirements of the flour have increased (not enough water in the mix, and therefore the contents of the mixing bowl use up the allotted energy at a faster rate). This may happen for a number of reasons, e.g. when the baker is using new harvest flour or a flour specification change has gone unnoticed. Extra water can be added to restore the time taken to impart the work input under the standard conditions stated and so restore the dough consistency to the required level. If the recorded time to reach a constant work input increases, then this indicates that the dough has become softer and less water would have to be added to subsequent doughs to restore them to the standard conditions.

In certain products such as pan breads or baguettes the higher water levels are required, whilst in others such as freestanding breads the levels are lower. In either case the aim is to obtain the dough consistency most appropriate for the subsequent processing of the dough after mixing in order to achieve the product quality sought. In modern mixing machines, automatic dough consistency control devices are available. It is advisable to have standard conditions for each product type and to monitor mixing so that such changes as described here are picked up quickly.

Flour type

As discussed above, the use of the CBP often permits a reduction in flour strength of approximately 1% of protein. However, it is not essential to use a lower-protein flour in the manufacture of bread by the CBP. Typical specifications for UK CBP-type flours are given in Table 4.1 (see Chapter 4). In the recipes given in this chapter the flour would normally be a CBP type, although a bakers' grade flour may be used. The precise flour specification should be arrived at by discussion between flour miller and baker.

Mixing conditions

Each product type will have appropriate mixing conditions, depending on flour type and strength and dough consistency requirements for further processing. As a general rule of thumb, for a flour of 10–12% protein, work input should be 11 W h/kg (5 W h/lb) imparted over 2–4 min. In practice, the operator multiplies the dough weight by 11 for kilograms or by 5 for pounds and sets this value on the watt–hour meter; the meter then allows the machine to run until the watt–hours have been used up. For flours of higher strength the work input should be proportionally higher (e.g. for a 14% protein flour the watt–hours should be $14/11 \times 11 = 14$ W h/kg). However, in the case of stronger flours the temperature rise in the mixer will be greater and so some form of cooling (cooling jacket) needs to be employed to ensure dough of the correct consistency for processing.

If the machine is capable of mixing with different mixer headspace pressures, then many attributes of bread quality may be adjusted by this method rather than through formulation or other process changes. In the recipes that follow, values and timings for mixing under different conditions are suggested.

As doughs made by the CBP are generally warmer and also contain more yeast than bulk fermented doughs, it is advisable only to mix doughs of a size that can be processed in a 10–15 min period. This is particularly true of products requiring a fine cell structure. The timing can be relaxed to 20–30 min where a more open structure, e.g. French sticks (baguettes), is required. Where there are specific mixing conditions for a product, these are detailed under the relevant product.

In the recipes below, the improver contains fat, soya flour, ascorbic acid and enzyme-active materials in the proportions listed in the general improver formulation given above. Where additional ingredients may be required in the improver, they have been noted in the recipe.

Salt and yeast levels

At the time of writing, there is a trend for salt levels to fall in bread recipes in response to concerns about the level of salt in the diet of the UK consumer. In the recipes which follow a range of commonly used salt levels from around the world are included by way of examples.

If salt levels are reduced, then it is common practice to adjust yeast levels downwards to help to control fermentation in the prover. Ranges for yeast levels have not been given to avoid over-complication which will inevitably arise when trying to take into account different yeast types, different proof volumes and recipe control factors such as salt.

Recipes

The recipes, improver formulations and processing variations are intended only as a guide. Local variations required to the recipe are inevitable.

Standard bread: semi-automatic plant (family bakery)

Table 9.3

Ingredient	Bakers' per cent (%)	Weight (kg)
Flour (CBP or bakers' grade)	100	128.0
Yeast	2.1	2.7
Salt	1.7–2.1	2.2–2.7
Water	61.4	78.6
Improver	1.5	1.9

Mixing options and notes
1   Mix at atmospheric pressure followed by a delayed partial vacuum, or use the full mix time at a partial vacuum which gives an increasingly finer cell structure. The latter will reduce ascorbic-acid-assisted oxidation and so may give lower bread volume (see Chapter 5).
2   A CBP commercial improver should contain fat, ascorbic acid and fungal *alpha*-amylase on a carrier such as soya or wheat flour.
3   Yeast levels may be reduced if lower salt levels are used.

Standard bread: plant production (Fig. 9.1)

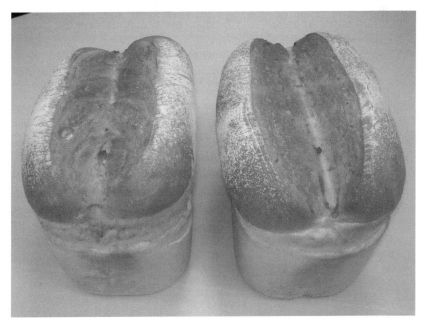

Fig. 9.1 Farmhouse bread.

Table 9.4

Ingredient	Bakers' per cent (%)	Weight (kg)
Flour (CBP or bakers' grade)	100	128
Yeast	2.1	2.7
Salt	1.7–2.1	2.2–2.7
Water	60.7	77.7
Improver	1.5	1.9

Mixing options and notes
1   Mix at atmospheric pressure followed by a delayed partial vacuum, or use the full mix time at a partial vacuum which gives an increasingly finer cell structure. The same mixing conditions will reduce ascorbic-acid-assisted oxidation and so may give lower bread volume (see Chapter 5).
2   Pressure–vacuum mixing provides an option for maintaining bread volume while achieving a fine uniform cell structure, e.g. 1.5 bar for 50% of energy (or time), plus 0.34 bar for the remainder. These conditions give the close and uniform structure typically required for this product.

3   A CBP commercial-powder-based improver should contain fat, ascor-
    bic acid and fungal *alpha*-amylase on a carrier such as soya or wheat
    flour.
4   A liquid-based improver may be used but should contain the essential
    functional ingredients.
5   Yeast levels may be reduced if lower salt levels are used.

High specific-volume bread

High specific-volume breads are often aimed at the weight- or health-conscious consumer. The recipe which follows is a generic form but branded products are available.

Table 9.5

Ingredient	Bakers' per cent (%)	Weight (kg)
Flour (bakers' grade)	100	128
Yeast	3	3.8
Salt	1.7–2	2.2–2.6
Water	62	79.4
Gluten*	0.7	0.9
Fat*	0.7	0.9
Ascorbic acid*	0.02	0.03

* These ingredients may be replaced with a suitable improver at 1.5–3.0% of flour weight.

Mixing options and notes
1 Pressure–vacuum mixing provides an option for maintaining bread volume while achieving a fine uniform cell structure, e.g. 1.5 bar for 50% of energy (or time), plus 0.34 bar for the remainder. These conditions give the high volume, close and uniform structure typically required for this product.
2 Enzymes may also be added to help to improve gas retention but levels of addition should not be so high as to make the crumb too soft and susceptible to slicer damage.

Oven-bottom and crusty bread

Table 9.6

Ingredient	Bakers' per cent (%)	Weight (kg)
Flour (CBP or bakers' grade)	100	128
Yeast	2.3	2.9
Salt	1.8–2.1	2.3–2.7
Water	53.0–57.0	67.8–73.0
Improver	1.5	1.9

This recipe is suitable for cottage loaves, tiger/leopard skin breads and Coburgs.

Mixing options and notes
1   Mix at atmospheric pressure followed by a delayed positive pressure, or use pressure throughout mixing which gives a progressively more open cell structure in products.
2   Added water levels may be reduced in order to restrict flow in the prover.

Scottish batch bread

Table 9.7

Ingredient	Bakers' per cent (%)	Weight (kg)
Flour* (CBP or bakers' grade)	100	128
Yeast	1.1–2.1	1.4–2.7
Salt	1.9–2.1	2.4–2.7
Fat+	0.7	0.9
Water	60.7	77.7
Ascorbic acid+	0.02	0.03

* Good quality breadmaking flour with low *alpha*-amylase activity.
+ These ingredients may be replaced with a suitable improver at 0.5–1.0% of flour weight.

Mixing options and notes
1 Mix at atmospheric pressure to the required work input.
2 Proof time may need to be longer if lower yeast and higher salt levels are used.

Irish batch bread

Table 9.8

Ingredient	Bakers' per cent (%)	Weight (kg)
Flour* (CBP or bakers' grade)	100	128
Yeast	2.1	2.7
Salt	1.8	2.3
Fat⁺	0.7	0.9
Water	51.8	66.3
Ascorbic acid⁺	0.02	0.03

* Use breadmaking flour with low *alpha*-amylase activity.
+ These ingredients may be replaced with a suitable improver at 0.63–1.0% of flour weight.

Mixing options and notes
1   Mix at atmospheric pressure to the required work input.

Brown bread

Table 9.9

Ingredient	Bakers' per cent (%)	Weight (kg)
Flour (CBP or bakers' grade)	100	128
Bran	10–15	12.8–19.2
Yeast	2.3	2.9
Salt	1.9–2.1	2.4–2.7
Fat*	1.4	1.8
Water	65–67	83.2–85.8
Ascorbic acid*	0.02	0.03

* These ingredients may be replaced with a suitable improver at 1.5% of flour weight.

Mixing options and notes
1   Mix at atmospheric pressure to the required work input.
2   The improver for brown bread should contain a higher proportion of fat (1.4%) as this will improve the volume.
3   Large bran flakes will give better crumb appearance than fine bran particles.

Vienna rolls and batons

Table 9.10

Ingredient	Bakers' per cent (%)	Weight (kg)
Flour (CBP or bakers' grade)	100	128
Yeast	3.1–3.6	4.0–4.6
Salt	1.6	2.1
Water	55.4	70.9
Fat*	1.4–1.8	1.8–2.3
Ascorbic acid*	0.02	0.03
Enzyme active malt flour*	0.4–0.5	0.5–0.6

* These ingredients may be replaced with a suitable improver at 2% of flour weight. Improver may typically contain fat at 1.4, enzyme-active malt flour at 0.4, ascorbic acid at 0.02 and soya flour or some other base flour.

Mixing options and notes
1   Vienna rolls and batons normally require a slightly stronger flour than used with pan breads.
2   The crumb cell structure of these products is fairly open and relatively random in nature. Such products can be produced using the CBP either by resting the dough after mixing for a long enough period for the yeast to start fermenting and expand the bubble structure in the dough (typically up to 20 min), or by mixing the dough under pressure for the whole or majority of the mixing period (see Chapter 5). In either case the longer resting period post-mixing is beneficial to an open random structure. Where the dough moulding happens within a few seconds after dividing as is the case in automatic roll plants, the dough can be rested in bulk after mixing for 10–15 min before continuing to the divider. With semi-automatic or hand moulding, an intermediate proof of 15–20 min is essential between first and second moulds. Processing can then continue as normal.
3   The product should be steamed within a few minutes of entering the oven. The moist atmosphere encourages the gelatinisation of the starches near the surface of the product and causes the crust to be shiny. Baking should continue in this atmosphere until the dough is fully expanded. The damper can then be pulled to allow baking to be completed in a dry atmosphere to encourage crust formation.

French sticks (Fig. 9.2)

Fig. 9.2   French sticks showing different cutting techniques.

There are two main types of product in the UK that are called 'French sticks'.

The first is the baguette, which is 70 cm long, scaled at 350 g of dough for a baked weight of 250 g. It has seven cuts along its length. The colour of the crust penetrates about 3 mm into the crumb and the hard crisp crust is sharp enough to cut the mouth when eaten.

The second is the pain Parisien. This is made from 550 g of dough, has a baked weight of 400 g, is 70 cm long and has five cuts on the top crust. In both cases the crumb is very open and random.

The French stick that is purchased from the UK in-store bakery is a cross between these two: 400 g baked weight and seven cuts along the top crust.

The typical French bread structure can be achieved using the CBP by the correct dough development, gentle dough manipulation and adequate dough gassing.

Table 9.11

Ingredient	Bakers' per cent (%)	Weight (kg)
Flour (CBP or bakers' grade)	100	128
Yeast	3	3.8
Salt	2	2.6
Water	66	84.5
Emulsifier*	0.3	0.4
Ascorbic acid*	0.02	0.03

* These ingredients may be replaced with a suitable improver at 1.5–2% of flour weight. In the manufacture of baguettes the emulsifier may be lecithin.

Mixing options and notes
1   Mix under a positive pressure, e.g. 1.5 bar, throughout to the required work input (11 Wh/kg) and a dough temperature of 31 °C.
2   If mixed under atmospheric pressure, the first proof time may be extended to 20 min to allow yeast to start working and to promote the large open-cell structure required by this product.
3   Final proof temperatures may be lower than for pan bread, e.g. 21 °C, with a proof time of 80 min, or a proof temperature of 43 °C with a proof time of 40 min may be used.
4   The product should be steamed within a few minutes of entering the oven. The moist atmosphere encourages the gelatinisation of the starches near the surface of the product and causes the crust to be shiny. Baking should continue in this atmosphere until the dough is fully expanded. The damper can then be pulled to allow baking to be completed in a dry atmosphere to encourage crust formation.

Soft round and finger rolls (Fig. 9.3)

Fig. 9.3   Soft rolls.

Table 9.12

Ingredient	Bakers' per cent (%)	Weight (kg)
Flour (bakers' grade)	100	128
Yeast	3.1	4.0
Salt	1.8	2.3
Water	61.4	78.6
Fat*	2.2	2.8
Ascorbic acid*	0.02	0.03
Milk powder	1.3	1.7
Sugar	1.6	2.1

* These ingredients may be replaced with a suitable improver at 2–2.5% of flour weight.

Mixing options and notes
1   Mix at atmospheric pressure to the required work input.
2   Milk powder should be heat treated.

Baps or plain teacakes

Table 9.13

Ingredient	Bakers' per cent (%)	Weight (kg)
Flour (bakers' grade)	100	100
Yeast	3.2	4.1
Salt	1.8	2.3
Water	57.1	73.1
Fat	10.7	13.7
Ascorbic acid	0.02	0.03

Mixing options and notes
1 Mix at atmospheric pressure to the required work input.
2 It is recommended that the improver for baps and plain teacakes contains 10.7% fat and 0.02% ascorbic acid.

Reduced-energy rolls

Table 9.14

Ingredient	Bakers' per cent (%)	Weight (kg)
Flour (bakers' grade)	100	128
Yeast	2.5	3.2
Salt	1.8	2.3
Water	168	215
Fat*	0.7	0.9
Ascorbic acid*	0.02	0.03
Calcium proprionate*	0.1	0.1
Sodium carboxymethyl cellulose	12.0	15.4

* These ingredients may be replaced with a suitable improver at 1.5–2.5% of flour weight.
+ May be replaced by other fibres.

Mixing options and notes
1   Mix at atmospheric pressure to the required work input.
2   Dough has a characteristic stickiness and it will be necessary to mould gently, e.g. by hand.
3   Scale at 56 g (baked weight, 48 g).
4   Give a first proof of 10 min, final mould and then prove for 45 min at 40 °C with 80% relative humidity.
5   Bake at 220 °C for 18 min.
6   Total energy per 100 g of baked roll is about 140 kcal or about 584 kJ.

Currant buns

Table 9.15

Ingredient	Bakers' per cent (%)	Weight (kg)
Flour (CBP or bakers' grade)	100	128
Yeast	5.4	6.9
Salt	0.7	0.9
Water	41.1	52.6
Fat	10.7	13.7
Milk powder	2.9	3.7
Sugar	10.7	13.7
Frozen egg	17.9	22.9
Ascorbic acid	0.02	0.03
Currants	28.6	36.6
Sultanas	28.6	36.6
Peel	5.4	6.9

Mixing options and notes
1   Mix the base (unfruited) dough at atmospheric pressure to the required work input.
2   The intense mixing of doughs in CBP causes severe break-up of dried fruit. Consequently the addition of the fruit should be made during the final stages of mixing. This addition is possible using machines with specially designed lids or special devices for adding fruit automatically. Alternatively, transfer base dough to a separate slow-speed mixer and blend in fruit.
3   When calculating the required watt–hour input mixing requirement for the dough, the weight of the fruit should be excluded.

Hot cross buns

Table 9.16

Ingredient	Bakers' per cent (%)	Weight (kg)
Flour (CBP or bakers' grade)	100	128
Yeast	8.9	11.4
Salt	0.8	1.0
Water	46.4	59.4
Fat	14.3	18.3
Milk powder	3.1	4.0
Sugar	14.3	18.3
Frozen egg	12.1	15.5
Ascorbic acid	0.02	0.03
Currants	14.3	18.3
Sultanas	7.1	9.1
Peel	0.7	0.9
Bun spice, as required		

Mixing options and notes
1   Mix the base (unfruited) dough at atmospheric pressure to the required work input.
2   The intense mixing of doughs in CBP causes severe break-up of dried fruit. Consequently the addition of the fruit should be made during the final stages of mixing. This addition is possible using machines with specially designed lids or special devices for adding fruit automatically. Alternatively, transfer base dough to a separate slow-speed mixer and blend in fruit.
3   When calculating the required watt–hour input mixing requirement for the dough, the weight of the fruit should be excluded.
4   Spices have antimicrobial properties which can have a harmful effect on yeast cells and on fermentation. A retarding effect can be seen. It is advisable to keep the yeast and spice apart until dough mixing begins. Care should be taken on the selection of spices since, if too high a level is used or a long time is taken to work the dough off, the dough may break down.

Hot cross bun paste

Table 9.17

Ingredient	Bakers' per cent (%)	Weight (kg)
Flour	100	128
Fat	20	25.6
Milk powder	12.5	16
Water	100	128
Baking powder	0.6	0.8

Mixing options and notes
1   The paste is mixed to a smooth batter and is piped on to the buns immediately after proving and before baking.

Bun glaze

Table 9.18

Ingredient	Amount (%)	Weight (kg)
Egg	50	50
Sugar	25	25
Water	25	25

Mixing options and notes
1   The ingredients are whisked together thoroughly. The glaze is brushed
over the buns immediately after baking.

Swiss buns

Table 9.19

Ingredient	Bakers' per cent (%)	Weight (kg)
Flour (CBP or bakers' grade)	100	128
Yeast	4.3	5.5
Salt	0.7	0.9
Water	58.9	75.4
Fat	7.1	9.1
Milk powder	2.9	3.7
Sugar	8	10.2
Ascorbic acid	0.02	0.03

This recipe is suitable for Devon splits, cookies and Parisian buns.

Mixing options and notes
1   Mix at atmospheric pressure to the required work input.

Doughnuts

Table 9.20

Ingredient	Bakers' per cent (%)	Weight (kg)
Flour (CBP or bakers' grade)	100	128
Yeast	8.9	11.4
Salt	0.8	1.0
Water	46.4	59.4
Fat	8.9	11.4
Milk powder	3.1	4.0
Sugar	10.7	13.7
Frozen egg	8	10.2
Ascorbic acid	0.02	0.03

Mixing options and notes
1   Mix at atmospheric pressure to the required work input.
2   After proof, the doughnuts are fried in oil at 182–193 °C (360–380 °F).

Tea bread and bun rings

Table 9.21

Ingredient	Bakers' per cent (%)	Weight (kg)
Flour (CBP or bakers' grade)	100	128
Yeast	5.4	6.9
Salt	0.7	0.9
Water	41.1	52.6
Fat	10.7	13.7
Milk powder	2.9	3.7
Sugar	10.7	13.7
Frozen egg	17.9	22.9
Ascorbic acid	0.02	0.03
Currants	28.6	36.6
Sultanas	28.6	36.6
Peel	5.4	6.9

Mixing options and notes
1   Mix the base (unfruited) dough at atmospheric pressure to the required work input.
2   The intense mixing of doughs in CBP causes severe break-up of dried fruit. Consequently the addition of the fruit should be made during the final stages of mixing. This addition is possible using machines with specially designed lids or special devices for adding fruit automatically. Alternatively, transfer base dough to a separate slow-speed mixer and blend in fruit.
3   When calculating the required watt–hour input mixing requirement for the dough, the weight of the fruit should be excluded.

Fruited teacakes

Table 9.22

Ingredient	Bakers' per cent (%)	Weight (kg)
Flour (CBP or bakers' grade)	100	128
Yeast	6.3	8.1
Salt	1.3	1.7
Water	53.6	68.6
Fat	8.9	11.4
Milk powder	2.7	3.5
Sugar	8.9	11.4
Ascorbic acid	0.02	0.03
Currants	18.8	24.1
Peel	2.7	3.5

Mixing options and notes
1   Mix the base (unfruited) dough at atmospheric pressure to the required work input.
2   The intense mixing of doughs in CBP causes severe break-up of dried fruit. Consequently the addition of the fruit should be made during the final stages of mixing. This addition is possible using machines with specially designed lids or special devices for adding fruit automatically. Alternatively transfer base dough to a separate slow-speed mixer and blend in fruit.
3   When calculating the required watt–hour input mixing requirement for the dough, the weight of the fruit should be excluded.

Malt bread

Table 9.23

Ingredient	Bakers' per cent (%)	Weight (kg)
Flour (CBP or bakers' grade)	100	128
Malt flour (low diastatic)	7.7	9.9
Yeast	2.3	2.9
Salt	1.1	1.4
Water	61.1	78.2
Fat	1.1	1.4
Ascorbic acid	0.02	0.03
Treacle	5.0	6.4

Mixing options and notes
1 Mix at atmospheric pressure to the required work input, usually lower than that for standard breads, e.g. 7 W h/kg, with final dough temperature of 31 °C (87 °F).
2 After mixing, the dough should be scaled, moulded, rested for 6–10 min, remoulded and panned.
3 Final proof will be 45–50 min at 43 °C (110 °F).
4 Baking time for a 400 g loaf will be about 45 min at 226 °C (440 °F).

Farmhouse rolls

Table 9.24

Ingredient	Bakers' per cent (%)	Weight (kg)
Flour (CBP or bakers' grade)	100	128
Yeast	5.4	6.9
Salt	1.8	2.3
Water	57.1	73.1
Fat*	1.6	2.1
Milk powder	2.2	2.8
Ascorbic acid*	0.02	0.03

* These ingredients may be replaced with a suitable improver at 2–2.5% of flour weight.

Mixing options and notes
1 Mix at atmospheric pressure to the required work input.
2 The milk powder should be fully heat treated.
3 Scale dough pieces at 300 g. After first moulding and intermediate proof, give final mould and place dough pieces about 2.5 cm (1 in) apart on baking sheets, lightly greasing between dough pieces.
4 Prove until dough pieces touch before baking.

Brown farls

Table 9.25

Ingredient	Bakers' per cent (%)	Weight (kg)
Flour (brown)	100	128
Yeast	2.2	2.8
Salt	2.1	2.7
Water	60.7	77.7
Fat*	1.4	1.8
Dark malt flour	1.4	1.8
Ascorbic acid*	0.02	0.03

* These ingredients may be replaced by an improver added at a level of 1–2% flour weight. Concentrates containing the malt flour may also be used at higher levels of addition.

Mixing options and notes
1   Mix at atmospheric pressure to a lower work input than for bread doughs, e.g. 9 W h/kg dough.
2   Scale dough pieces at 900 g, mould round and give 10 min rest period.
3   Pin out to 20 cm (8 in) diameter and cut into four.
4   Give full proof and then dust lightly with coarse brown flour or bran flakes before baking.

Hamburger rolls: UK style

Table 9.26

Ingredient	Bakers' per cent (%)	Weight (kg)
Flour (CBP or bakers' grade)	100	128
Yeast	4	5.1
Salt	1.8	2.3
Water	55.4–62.2	70.9–76.6
Fat	5.0–5.4	6.4–6.9
Milk powder	1.6–2.0	2.1–2.6
Sugar	3.0–3.6	3.8–4.6
Ascorbic acid*	0.02	0.03

* May be added as part of an improver.

Mixing options and notes
1   Mix at 1.5 bar for 50% of time, and at 0.34 bar for the remainder of the time. These conditions give the close and uniform structure typical of this product.
2   Alternatively, mix at atmospheric pressure to the required work input for a more open cell structure.
3   Scale dough into 60 g pieces and give a first proof of 10 min before continuing processing.
4   Crust colour is increased by raising milk powder and sugar contents. Softness is improved by increasing the fat level, and tenderness is improved by increasing both fat and sugar.

Hamburger rolls: US style

Table 9.27

Ingredient	Bakers' per cent (%)	Weight (kg)
Flour (CBP or bakers' grade)	100	128
Yeast	3	3.8
Salt	1.8	2.3
Water	55.4	70.9
Fat	3–5	3.8–6.4
Milk powder	1.6	2.1
Sugar	5–10	6.4–12.8
Ascorbic acid*	0.02	0.03

* May be added as part of an improver. In some parts of the world, other oxidants may be present.

Mixing options and notes
1   Mix at 1.5 bar for 50% of time, and at 0.34 bar for the remainder of the time. These conditions give the close and uniform structure typical of this product.

Brown teacakes

Table 9.28

Ingredient	Bakers' per cent (%)	Weight (kg)
Flour (brown)	100	128
Yeast	3.6	4.6
Salt	1.8	2.3
Water	60.7	77.7
Fat	7.1	9.1
Milk powder	2.7	3.5
Ascorbic acid*	0.02	0.03

* May be added as part of an improver.

Mixing options and notes
1   Mix at atmospheric pressure to the required work input.

Irish turnovers or grinders

Table 9.29

Ingredient	Bakers' per cent (%)	Weight (kg)
Flour (CBP or bakers' grade)	100	128
Yeast	2.1	2.7
Salt	1.8	2.3
Water	51.8	66.3
Fat*	0.7	0.9
Ascorbic acid*	0.02	0.03

* These ingredients may be replaced with a suitable improver at 1.0% of flour weight.

Mixing options and notes
1   Mix at atmospheric pressure to the required work input.

Irish soda bread (with yeast)

Table 9.30

Ingredient	Bakers' per cent (%)	Weight (kg)
Flour (CBP or bakers' grade)	100	128
Yeast	3.6	4.6
Salt	1.6	2.1
Water	60.7	77.7
Fat	3.6	4.6
Malt flour*	0.7	0.9
Baking powder	3.1	4.0
Milk powder	1.6	2.1
Ascorbic acid*	0.02	0.03

* May be added as part of an improver.

Mixing options and notes
1   Mix at atmospheric pressure to the required work input.
2   After scaling and intermediate proof, allow to prove for about 40 min at 30–35 °C.
3   Score lightly with a knife and bake for 35 min at 232 °C (450 °F).

Milk bread: containing whole-milk solids

Table 9.31

Ingredient	Bakers' per cent (%)	Weight (kg)
Flour (CBP or bakers' grade)	100	128
Yeast	2.1	2.7
Salt	2.1	2.7
Water	18.6	23.8
Fat*	1.8	2.3
Milk	50.7	64.9
Ascorbic acid*	0.02	0.03

* May be added as part of an improver used at 2% flour weight.

Mixing options and notes
1   Mix at atmospheric pressure to the required work input.

Milk bread: containing skimmed-milk solids

Table 9.32

Ingredient	Bakers' per cent (%)	Weight (kg)
Flour (CBP or bakers' grade)	100	128
Yeast	2.1	2.7
Salt	1.8	2.3
Water	62.5	80
Fat*	0.7	0.9
Skimmed-milk powder	6.1	7.8
Ascorbic acid*	0.02	0.03

* May be added as part of an improver used at 1–2% flour weight.

Mixing options and notes
1   Skimmed-milk powder should be heat treated.
2   Mix at atmospheric pressure to the required work input.

Wholemeal or wholewheat bread

Table 9.33

Ingredient	Bakers' per cent (%)	Weight (kg)
Wholemeal flour	100	128
Yeast	2.7	3.5
Salt	1.8	2.3
Water	62.5	80
Fat*	1.8	2.3
Ascorbic acid*	0.02	0.03

* These ingredients may be replaced with a suitable improver at 2.0% of flour weight.

Mixing options and notes
1 The presence of high levels of bran in wholemeal flours reduces dough gas retention and therefore mixing under partial vacuum is likely to lead to lower bread volume.
2 Mix at atmospheric pressure to the required energy input.
3 Alternatively mix under positive pressure (up to 1.5 bar) for half the energy input and complete mixing at atmospheric pressure.
4 The cell structure of wholemeal bread will tend to be more open than that of white bread of the same specific volume.
5 After final proof, bake for 30 min at 227 °C (440 °F).

Mixed-grain bread

Table 9.34

Ingredient	Bakers' per cent (%)	Weight (kg)
Mixed-grain flour	100	128
Yeast	2.1–2.3	2.7–2.9
Salt	1.8–2.1	2.3–2.7
Water	60.7	77.7
Fat*	0.71	0.9
Ascorbic acid*	0.02	0.03

* These ingredients may be replaced with a suitable improver at 2.0% of flour weight.

Mixing options and notes
1   The presence of high levels of bran and grains in mixed-grain flours reduces dough gas retention and therefore mixing under partial vacuum is likely to lead to lower bread volume.
2   Mix at atmospheric pressure to the required energy input. Alternatively, mix under positive pressure (up to 1.5 bar) for half the energy input and complete mixing at atmospheric pressure.

Rye bread

Table 9.35

Ingredient	Bakers' per cent (%)	Weight (kg)
Flour	100	100.0
Yeast	3	2.1
Salt	1.6	2.1
Water	60.7	60.7
Fat	0.71	1.5
Ascorbic acid	0.02	0.03

Mixing options and notes
1   In the case of all-rye flours the work input during mixing will not be critical and there will be little or no advantage gained from the addition of ascorbic acid.
2   In the case of mixed rye and wheat flours, the work input during mixing can be dropped to 4–7 W h/kg as the mixed flours respond very little to work variation.
3   Mix at atmospheric pressure.
4   If the authentic 'sour' flavour of the rye bed is desired, it is necessary to prepare a 'sour' about 12 h prior to dough mixing. The 'sour' should be added at the rate of 6% of flour weight.

Sponge and dough, including North American pan breads
Breads made by the 'sponge and dough' or 'flour brew' methods can also
be made using the CBP. The sponge containing a proportion of the flour,
yeast and water is mixed and held in a tank for a fixed period of time
(depending on the flavour required). This sponge is then added to the recipe
as an 'additional' ingredient and the dough is mixed as in normal CBP
processing.

Sponge recipe for 16h sponge

Table 9.36

Ingredient	Bakers' per cent (%)	Weight (g)
Flour (CBP or bakers' grade)	100	2500.0
Yeast	0.8	20.0
Salt	1	25.0
Water	56	1400.0

Mixing options and notes
1   Mix ingredients to a clear dough using any kind of mixing machine and
    ferment sponge at 21 °C (70 °F) for 16h.

Recipe for final dough

Table 9.37

Ingredient	Bakers' per cent (%)	Weight (g)
Flour (CBP or bakers' grade)	100	7500.0
Yeast	2.4	180.0
Salt	2.4	180.0
Water	61.3	4600.0
Improver	1.5	110.0
Sponge	52.6	3945

Table 9.38

Ingredient	Amount (% of total flour weight)	
	Sponge	Dough
Flour	25.0	75.0
Yeast	0.2	1.8
Salt	0.25	1.8
Water	14.0	46.00
Improver	0	1.1
		+sponge

Mixing options and notes
1   Mix at atmospheric pressure to the required energy input.
2   Finer crumb cell structures may be obtained with initial mixing at 1.5 bar and then moving to 0.5 bar halfway through the mixing cycle.
3   The weight of the sponge should be discounted for the purpose of calculating the total expenditure of energy in watt hours because it has in effect been 'developed' by the fermentation period.
4   The ingredients should be mechanically developed with an energy expenditure of 11–17 Wh/kg of dough (depending on flour strength) excluding the sponge. The cell structure of the bread made by sponge and dough methods tend to be more open than those produced by the conventional method. The saving in yeast will be quite small. A standard CBP recipe uses 215 g of yeast per 10 kg of flour, whilst this recipe uses 200 g.
5   As there is less yeast in this recipe, the proof time may need to be increased to maintain proof and loaf volume. The usual relationship between yeast and salt will still apply so that a reduction in yeast may be accompanied by a reduction in salt level.
6   With respect to the choice of flour, use bakers' grade or stronger flour. The stronger flour could be used for the sponge stage where greater tolerance is required and baker's grade for the dough stage.
7   The optional ingredients are as follows: 0.5–1% full-fat enzyme-active soya flour on flour weight will improve crumb colour, volume and softness; 0.2% malt flour will improve crust colour, volume and softness; 2% skimmed-milk powder will improve crust colour and taste; extra fat will give a tender crust and a softer crumb.
8   North American sponges tend to be held for shorter times, typically 4 h and the dough recipe will contain milk solids and fat.

Australian and New Zealand breads

Australia and New Zealand embraced the CBP although they tend to refer to it as mechanical dough development (MDD). In essence the two processes are very similar. Mixing time rather than work (energy) input is the measured controlled parameter. Australian flours have higher protein levels and consequently the work input needs to be greater (typically around 14 Wh/kg) and the resulting temperature rise needs to be kept under control.

White bread

Table 9.39

Ingredient	Bakers' per cent (%)	Weight (kg)
Flour (CBP or bakers' grade)	100	128.0
Yeast	4.0	5.1
Salt	1.8	2.3
Water	61.0	78.1
Improver	1.0	1.3
Vegetable oil	1.0	1.3
Softener	0.4	0.5
Soya flour	0.5	0.6
Calcium acetate	0.3	0.4

Mixing options and notes
1 Mix at atmospheric pressure followed by delayed partial vacuum or use the full mix time at a partial vacuum which gives an increasingly finer cell structure. The same mixing changes will reduce ascorbic-acid-assisted oxidation and so may give lower bread volume.
2 Generally the improver may contain flour, soya flour, emulsifiers and enzymes.
3 Generally the softener contains a blend of enzymes for freshness.
4 Improver contains ascorbic acid at about 120 ppm.

Wholemeal bread

Table 9.40

Ingredient	Bakers' per cent (%)	Weight (kg)
Flour (CBP or bakers' grade)	85	109
Mealmix	15	19
Yeast	4.0	5.1
Salt	1.8	2.3
Water	68.0	87.0
Improver	1.3	1.7
Vegetable oil	1.0	1.3
Softener	0.4	0.5
Gluten	2.0	2.6
Calcium acetate	0.3	0.4
Dark malt flour*	0.5	0.6

* Optional, for colour only.

Mixing options and notes
1   The presence of high levels of bran in wholemeal flours reduces dough gas retention and therefore mixing under partial vacuum is likely to lead to lower bread volume.
2   Mix at atmospheric pressure to the required energy input. Alternatively, mix under positive pressure (up to 1.5 bar) for half the energy input and complete mixing at atmospheric pressure.
3   The cell structure of wholemeal bread will tend to be more open than that of white bread of the same specific volume.
4   After final proof, bake for 30 min at 227 °C (440 °F).
5   Improver contains ascorbic acid at about 120 ppm.

Indian breads

Table 9.41

Ingredient	Bakers' per cent (%)	Weight (kg)
Flour	100	100.0
Yeast	2.1	2.1
Salt	2.1	2.1
Water	60.7	60.7
Improver	1.5	1.5
Sugar	Up to 30	Up to 30

Mixing options and notes
1 Mix at atmospheric pressure to the required work input or with a delayed partial vacuum.
2 Breads contain a proportion of sugar with the level of sugar being higher in the south than in the north of the country. Typical levels will start at 10% flour weight rising to as much as 30% in the south.

Chinese steamed bread (man-t'ou)
See the description given by Cauvain and Huang (1986).

Table 9.42

Ingredient	Bakers' per cent (%)	Weight (kg)
Untreated unbleached flour	100	100.0
Yeast	1	2.1
Water	As determined	60.7
Ascorbic acid	0.0075	

Mixing options and notes
1 Mix at atmospheric pressure to the required work input (11 W h/kg) with a dough temperature of 30.5 ± 1 °C.
2 The scaling weight is 100 g or 200 g, and the mould round.
3 No intermediate proof is required; prove at 43 °C and 85% relative humidity for no more than 15 min (100 g pieces) or 30 min (200 g pieces).
4 Steam for 15–20 min on wire mesh trays suspended in a steamer.

Pizza bases

Table 9.43

Ingredient	Bakers' per cent (%)	
	Rich recipe	Lean recipe
Flour (CBP or bakers' grade)	100	100
Yeast	7.0	6.5
Salt	2.0	2.0
Water	57	57
Fat*	1.0	0.7
Oil	8.0	0
Sugar	1.5	0
Skimmed-milk powder	2.5	0
Ascorbic acid*	0.003	0.003

* These ingredients may be included in the form of a CBP-type improver, which may or may not include sugar, fat and skimmed milk powder.

Mixing options and notes
1   Mix at atmospheric pressure to the required work input (11 W h/kg) with a dough temperature of 30.5 ± 1 °C.
2   Dough pieces should be moulded round, rested for 10 min and then pinned out to a circular shape (approximate thickness should be 4 mm).
3   Bases should be placed on a greased or silicone-lined trays and docked.
4   Prove at 43 °C and 80% relative humidity for 30–40 min until the dough has roughly tripled in its original thickness.
5   Bake at 204 °C (400 °F) for 8–10 min until bases turn a pale golden colour.

Retarding Chorleywood Bread Process doughs

Products mixed using the CBP can be successfully retarded provided that a few basic rules are adhered to (see Cauvain, 1998a, 1998c). They include the following:

1   The dough should be mixed to its normal temperature to achieve good gas-holding properties.
2   The dough should be scaled and processed as normal, certainly avoiding any delays.
3   Mixing and processing should not begin until the retarder temperature has reached the required low-level storage temperature.
4   It is important that the dough does not start to gas significantly before it is placed in the retarder as this may lead to an uneven cell structure in the final product.
5   The yeast level should be reduced by approximately 25% (depending on the length of the retard).
6   The settings for the retarder should allow for rapid cooling of the dough, without freezing its surface (e.g. $-4\,°C$), and for a long slow low-temperature proof (e.g. minimum of 4h at $21\,°C$).
7   This enables the dough to prove evenly throughout its structure rather than, in the case of a short sharp proof after retard, causing the outside to over-prove and the inside to under-prove.
8   Because enzymic activity continues in the dough during the retard phase, breads which have been retarded may have a darker colour than scratch-produced doughs.
9   The temptation to reduce baking time or temperature should be resisted as if the products are baked 'to colour', the inside may be under-baked.

Part-baked products
Part-baking is a process in which proved dough is baked just sufficiently to kill the yeast, to inactivate the enzymes and to set the product structure with the minimum of crust colouration and moisture loss (Collins *et al.*, 1971; Collins and Ford, 1985). The retailer or consumer completes the baking in the shop or home in order to provide freshly baked bread as close to the time of consumption as possible.

The CBP can be used to produce the dough for such products and the processing details for breads, rolls and French sticks is given below.

Part-baked bread

Table 9.44

Ingredient	Bakers' per cent (%)	Weight (kg)
Flour (CBP or bakers' grade)	100	128.0
Yeast	1.8–2.1	2.3–2.7
Salt	1.8	2.3
Water	61	78
Fat*	0.7	0.9
Ascorbic acid*	0.02	0.03

* May be added as part of an improver used at 1–2% flour weight. Propionic acid as a preservative may be added at 0.1% to assist mould-free or rope-free shelf life.

Mixing options and notes
1   Mix at atmospheric pressure to the required work input.
2   Prove dough to the same height as for a scratch product. Do not under-prove.
3   At low bake temperatures, dough continues to prove in the early stages of baking. To prevent over-proof from occurring, yeast levels may be reduced.
4   Bake to achieve a centre crumb temperature of 92–96 °C (198–204 °F) at a temperature of 150 °C (300 °F) for between 25 and 40 min depending on dough piece size and shape. To prevent ragged oven spring break, the steaming time can be increased.
5   Cool quickly for minimum moisture loss.
6   Bake off (second bake) at 280 °C for minimum time to impart colour to the crust.
7   Part-baked bread has a moisture content of approximately 41% compared with 37% for fully baked bread.

Part-baked rolls

Table 9.45

Ingredient	Bakers' per cent (%)	Weight (kg)
Flour (CBP or bakers' grade)	100	128.0
Yeast	3.0	3.8
Salt	1.8	2.3
Water	55	70
Fat*	2.2	2.8
DATA ester*	0.3	0.4
Gluten	2.0	2.6
Skimmed-milk powder	1.3	1.7
Sugar	1.6	2.1
Ascorbic acid*	0.02	0.03

* May be added as part of an improver used at 1–2% flour weight. Propionic acid as a preservative may be added at 0.1% to assist mould-free or rope-free shelf life. The improver may also contain fungal *alpha*-amylase.

Mixing options and notes
1   Mix at atmospheric pressure to the required work input.
2   Scale dough into 'heads' and rest for 10 min before dividing and moulding or pass through rollplant.
3   Prove dough at 43 °C and 80% relative humidity to the same height as for a scratch product. Low proof volume with light steaming during baking can be tolerated. Slightly more proof volume can be tolerated but with no steaming during baking.
4   Bake to achieve a centre crumb temperature of 92–96 °C (193–204 °F) at a temperature of 180 °C for 18 min. Products may be steamed for a few seconds (18 s) at the start of baking. Steaming for too long may cause wrinkling of the product on cooling.
5   Baking at 165 °C for 20 min will give a thicker crust.
6   Cool quickly for minimum moisture loss.
7   Products may be blast frozen to a core temperature of −10 °C and stored at −18 °C before baking off.
8   Rolls may be coated with starch paste before blast freezing or second bake to improve crust bloom.
9   Bake off (second bake) at 280 °C for 4 min to impart colour to the crust. Steam may be used to give a good crust bloom.

Part-baked French sticks

Table 9.46

Ingredient	Bakers' per cent (%)	Weight (kg)
Flour (CBP or bakers' grade)	100	128.0
Yeast	3.0	3.8
Salt	2.0	2.6
Water	60	76.8
Fat*	0.7	0.9
DATA ester*	0.3	0.4
Gluten	2.0	2.6
Ascorbic acid*	0.02	0.03

* May be added as part of an improver used at 1–2% flour weight. Propionic acid as a preservative may be added at 0.1% to assist mould-free or rope-free shelf life. The improver may also contain fungal *alpha*-amylase.

Mixing options and notes
1   Mix at atmospheric pressure to the required work input or at a positive pressure to create a more open cell structure.
2   Scale and rest for 20 min.
3   Prove dough at 32 °C, 70% relative humidity to the same height as for a scratch product (approximately 70 min). Low proof volume with light steaming can be tolerated. Slightly more proof volume can be tolerated but with no steaming during baking.
4   Bake to achieve a centre crumb temperature of 92–96 °C (193–204 °F) at a temperature of 180 °C for 20 min. Products may be steamed for at most 18 s. Excess steaming will cause collapse and wrinkling during cooling.
5   A higher temperature and a shorter bake time, e.g. 240 °C for 14 min, will also give acceptable quality with a light crust colour and can be used if the product is not to be frozen.
6   Cool quickly for minimum moisture loss.
7   Products may be blast frozen to a core temperature of –10 °C and stored at –18 °C before baking off.
8   Sticks may be coated with starch paste before blast freezing or second bake to improve crust bloom.
9   Bake off (second bake) at 280 °C for 6 min to impart colour to the crust. Steam may be used at bake off to give good crust bloom.

Starch-based coating for part-baked rolls and sticks

Table 9.47

Ingredient	Amount (parts)
Starch (wheat)	100
Gluten	16
Skimmed-milk powder	1.7
Salt	1.3
Fat	6.0
Water	2340

Mixing options and notes
1   Mix dry ingredients with 100 parts of water to form a thick paste.
2   Dilute with remaining water and heat until gelatinised.
3   Brush paste over cooled part-baked products before blast freezing.

Frozen unproved doughs

Doughs may be frozen after mixing by the CBP (Cauvain, 1981; Cauvain, 1998a). Important considerations include the following:

1 The one major adjustment that needs to be made in the recipe is to increase the yeast level to compensate for gassing losses during storage. For short storage periods, say 3–4 days, an increase in yeast level may be unnecessary. For a storage period of up to 30 days, the yeast level should be increased by 50%. For periods in excess of 30 days, an increase of 100% may be necessary.

2 An important aspect about the basic recipe (before adjustment of yeast) is that it must be capable of producing acceptable bread from the freshly prepared dough. Freezing and storing doughs does not improve bread quality and so it is essential to ensure that the choice of fat and composite improver used are compatible for the process.

3 In mixing the dough no changes are necessary to the process or to the dough temperature. For example, it is not advantageous to produce cold doughs from the mixer. In fact, it is better to produce a dough which is capable of retaining the gas in the harsh conditions during freezing and storage.

4 Doughs should be processed as quickly as possible in order to avoid excessive gas production, especially when higher than normal yeast levels have been used. To achieve more rapid processing in these circumstances it may be better to mix smaller quantities of dough than normal.

5 Dough pieces should be transferred to the freezing unit as quickly as possible after final moulding, again avoiding any delays to the process. The use of blast freezers, with their ability to freeze as rapidly as possible to the required temperature, gives best results.

6 The dough pieces should be wrapped in impermeable polyethylene bags and transferred to deep-freeze storage.

7 On a practical note, it is essential that the frozen pieces will fit in the pan or on the trays before defrosting as freezing fixes the dimensions of the dough piece. Some adjustment to final moulder settings or some restriction of movement may be necessary.

8 Frozen dough pieces are extremely hard and in consequence it may be advisable to ensure that adequate safety precautions are taken, e.g. protective head gear if overhead conveyors are used, protective footwear and insulated gloves.

9 A wide range of defrosting and proving combinations may be used with frozen dough depending on yeast level and product type. The lower yeast levels and cooler proof temperatures tend to give dough pieces that are more tolerant to variations in proof times. A gradual return to proof will benefit the dough rather than shorter sharp heating conditions during proof which will result in uneven product quality. Once

satisfactory defrosting and proving settings have been achieved, they should be adhered to.

10   Different bread varieties have different defrosting and proving requirements and so it will be difficult to defrost and prove combinations of different-sized products simultaneously in the same cabinet. Careful production planning or several defrosting and proving units will be required.

11   Defrosting, recovery and proof may be achieved in one continuous operation in commercially available equipment. In such equipment the maximum temperature required for any stage is pre-set together with the time for each part of the cycle. This facilitates gradual defrosting and proving of the dough pieces.

12   In a bakery which has a retarder (or retarder–prover) and a prover, the commercial continuous operation can be simulated. For example, to achieve this, dough pieces can be placed in a retarder set at 2–4 °C for 16 h during which time defrosting occurs. The dough pieces can then be transferred to a prover running at 21 °C for 1 h followed by 32 °C for 1 h and then 43 °C until proved. This will minimise the temperature differential within the dough pieces whilst gradually raising their temperature.

10

Applications of the Chorleywood Bread Process

Plant bakeries

The CBP is most closely associated with the large-scale mass production of bread in industrial plant bakeries. In part, this is because the CBP offers significant advantages for automation and control of a complex process while ensuring great consistency of final product, as discussed above. The principles used in modern CBP plant bakeries remain essentially the same as those first identified in the publications of 1961 although the detailed working practices may have changed significantly, if only because of a greater degree of automation and control. In plant bakeries the move has been to the automatic delivery of the ingredients, especially the flour and water, and more recently yeast and functional ingredients. Weighing devices have become more accurate through the use of load cells so that precise monitoring of the quantity of flour delivered to the mixer bowl is possible. In turn, this means that the water level delivered to the mixer can be based on the 'actual' flour weight delivered to the mixer and this gives the opportunity for greater consistency of doughs batch to batch. Mixing and dough development in the CBP remains a single operation lasting between 2 and 5 min to a defined energy input. Closer control of ingredient weights and their accurate recording permit adjustment of the mixing energy through PC control. Control of final dough temperature after mixing is achieved through measurement of ingredient temperatures and (short-time) historical data on final dough temperature and temperature rise for a given energy input. Control of cooled water to deliver the required water temperature to the mixer is now readily achieved automatically.

In some plant bakeries the delivery of the yeast and other functional ingredients may still be achieved by hand with the mixer stationary at 'half-tilt'. Improvers still remain a key component of bread made by the CBP and these have traditionally have been in the compressed form for yeast and the powder form for the improver. However, with improvements in metering technologies to dough mixers there are now greater opportunities for 'fluid' delivery of such materials. For example, cream yeast is now most commonly seen in plant bakeries and liquid-based improvers have also been introduced.

Cell structure control by changing the mixer headspace pressure remains a key component of the CBP in plant bakeries. The main control is still the application of a partial vacuum in order to arrive at a fine and uniform cell structure and to confer improved softness and crumb brightness considered desirable in many types of bread product, especially those destined for sandwich making. The introduction of the partial vacuum is automatically controlled in the plant bakery and is commonly delayed to some moment after the mixing cycle has begun. Some plant bakeries have adopted the pressure–vacuum mixer which offers greater flexibility in ascorbic acid oxidation in the dough and cell structure control in the final product. The advantage for the plant bakery is that it is able to make a wide range of bread types from a basic bread recipe with limited plant changes (Cauvain, 1994).

The CBP has always found its greatest application in the plant bakery where the high level of process control which can be achieved enables the large-scale production of standardised products. The CBP was quickly adopted in many countries throughout the world as discussed previously. While many would consider it to be a 'mature' baking technology, new users of the process are still arriving on the baking scene. Most often these new users are in the plant baking area having recognised the value of economies of scale and improved process control. Another driver which has led to the adoption of the CBP in areas where it has not been seen in the past has been the desire to produce the 'British sandwich loaf'. No other bread-making process can be used to reproduce the bread qualities concerned as many who have tried have discovered for themselves.

Craft bakeries

It is somewhat ironic that a breadmaking process originally envisaged for the small craft baker is rarely seen in that context in the UK today. Initially the uptake of the CBP by UK craft bakers in the 1960s was considerable and even through the 1970s there were still considerable numbers of users. As the numbers of practising craft bakers declined in the 1980s and 1990s, so did the proportion using the CBP. Many reasons have been given for this decline in use. In part, it is directly associated with decline in the absolute numbers of bakers in the UK. From its earliest introduction some of the fiercest critics of the CBP were the craft bakers who saw the loss of the bulk fermentation period as being synonymous with a loss of flavour and product quality in general; yet for many years the many small-scale users of the CBP were able to maintain their position in the market place.

The 'failure' of the CBP in the UK craft bakery is more likely to be associated with the nature of the process rather than with the products from it. The problems started with the fundamental lack of understanding of the nature of the energy–dough development relationship and the associated heat rise. Two of the key factors which were critical to its successful application were either misunderstood or ignored. The critical need for controlled work input has been discussed extensively above. In the bakery this needs to be achieved through the use of a watt–hour meter; however, such equipment costs money and often the bakers were reluctant to have one fitted, preferring instead to rely on control through a timer. Mixing to a fixed time inevitably leads to variations in dough development and subsequent bread quality. To some bakers it was simply too much effort to change the work input setting every time that they made a different-sized mixing; so they ignored this part of the process. The second problem for the small baker was the significant heat rise that occurred during mixing and the need for a low water temperature. In order to have sufficient cool water available, some form of water chilling was required. This equipment also cost money and was seen by many as an extravagance both in terms of initial purchase price and then in running costs.

The limited uptake of the CBP by the craft bakers for whom it was originally intended is based on a number of different factors. In part, it arises because, at the same time that the CBP was being introduced, other no-time breadmaking processes were also becoming available. One of note is the introduction of the ADD process (Brown, 1993). This process used 'chemical' development (based on an improver blend of ascorbic acid, potassium bromate and L-cysteine) to deliver a developed no-time dough. The appeal of ADD for the smaller baker was that it could be used with their existing, and generally low-speed, mixers and therefore did not need the capital investment required for application of the CBP. Since ADD was based on low-speed mixing, control of energy input was not required and it was easier to control dough temperature because of limited temperature rise in the low-speed low-friction mixing machines. All in all, ADD offered a more 'tolerant' process for bakeries where precise process control was more difficult to achieve. Eventually ADD would give way to spiral-mixed no-time doughs as first potassium bromate was withdrawn from use in the UK and the use of L-cysteine became uncommon, if only because of the high cost of the raw material.

The other reason for a limited uptake of the CBP by many smaller bakers was undoubtedly the strongly expressed beliefs by some craft bakers that the CBP produced an inferior loaf in terms of flavour. This belief was based on the need for a period of fermentation for the generation of 'flavour' in the breadcrumb. The elimination of the bulk fermentation period was seen by some as being directly responsible for a reduction of flavour in bread made by the CBP, even when presented with objective evidence to the contrary. The belief that the CBP produced bread with less flavour became one of the myths of the UK and other baking industries and in many cases resulted in a reluctance to undertake practical trials to try the process. In some ways the situation evolved that no 'self-respecting' craft baker would dare to operate the process because of potential peer pressure. It is an interesting point of speculation as to what may have happened if UK craft bakers had more readily grasped the commercial advantages offered by the CBP as was/is the case in some other parts of the world (e.g. see the discussion related to bread production in India in the section entitled International applications below).

In-store bakeries

To some extent the history of the application of the CBP within in-store bakeries mirrors that observed above for craft bakeries. In the UK, in-store bakeries began to develop within a retail environment at some time after the introduction and industrial acceptance of the CBP. By the time that in-store baking was being developed a number of craft bakeries were already successfully operating the CBP, which encouraged its adoption within the retail environment, as in many ways an in-store bakery resembles a smaller craft bakery.

When first introduced into in-store bakeries, CBP-type mixers were fitted with energy meters but with the passage of time they were abandoned in favour of the control of mixing using a timer. While it is true that the variation in mixing times for a given product dough mix of a given product size and with a given energy input was small, the move to use time alone to control dough mixing meant that variations in final product quality were inevitable. Variations in batch size would make a considerable contribution to product variability if time were used to control the mixing cycle.

The control of final dough temperature presented a familiar problem, not least because most in-store bakeries were not equipped with water chillers or, if they were, then the refrigerated water supply was inadequate for normal production levels. The lack of process control resulting from both the lack of using energy and water meters inevitably resulted in the in-store bakeries following the same path as many craft bakers, namely the abandonment of CBP in favour of using processes based on spiral-type mixers. It is ironic that some time after CBP was abandoned by in-store bakeries most have made the move to include water chillers and meters in order to control final dough temperatures and dough consistency. It is interesting to speculate what the fate of the CBP in in-store bakeries would have been if the application of the process had been strictly adhered to and if water chillers had been fitted as standard.

International applications

The international application of the CBP has been considerable since its launch in 1961. It has been estimated that the CBP has been used in 28 countries worldwide. The most significant application of the CBP has been in English-speaking countries. In Australia and New Zealand they prefer to consider that they are using MDD because they tend to use higher work input levels than the 11 W h/kg dough which is most commonly associated with the CBP. The need to use higher work input levels in Australia and New Zealand was recognised soon after the introduction of the CBP into those countries (Gould, 1998) and arose because of the greater strength of wheat flours used. However, such bakeries have developed an approach which allows them to produce acceptable bread and so the CBP, or MDD, remains the method which produces the largest proportion of bread in Australia and New Zealand. In common with the UK, the CBP tends to be found almost exclusively in plant bakeries.

Another country in which CBP dominates bread production is South Africa. The plant bakeries in that country also have to cope with wheat flours which achieve optimum bread quality (i.e. maximum bread volume) with energy inputs of over 11 W h/kg. Once again this is because of the strong wheat flours which are available. However, it is common for most bakeries to keep close to the work input level of 11 W h/kg dough. Dough temperature control is inevitably a problem in countries with high ambient temperatures, such as South Africa, and so the provision of chilled water or the use of ice in the mixer is practised.

One country with a warm climate in which the CBP has been used for some time is India. The process is used in plant and smaller bakeries; indeed, in some ways, many bakeries using the CBP in India resemble the craft bakeries of some years ago in the UK. Control of dough temperature is a significant problem in Indian bakeries where ambient temperatures may reach 40 °C or more. The inevitable consequence of raised dough temperatures is that dough becomes soft and sticky. In many bakeries this would present problems when the dough passes into the processing plant; in India the problems are less severe since hand processing of the dough is often seen. Wheat flour provided by Indian flour mills tends to be weaker than that seen in many countries, including the UK. This means that optimum work input levels are lower than seen in the UK and elsewhere and presents a part-solution to the temperature rise problem and chilled-water requirements. Typically, CBP-type mixers used in India will be set to run to a specific mixing time rather than to a defined work input.

Of the major English-speaking countries of the world the USA and Canada have been the exceptions when it came to the adoption of the CBP. Some bakeries did adopt the process and a couple of plants continue to operate it today. The strength of north American wheats was partly responsible for the apparent 'failure' of the CBP in North America and yet it is ironic that some UK bakeries use wheat flour grists with high proportions of North American wheats to produce their bread. Together with the greater flour strength there were potential problems of keeping dough temperatures to a sufficiently low level for subsequent dough processing. Bakeries in North America have never readily adapted to final dough temperatures as high as 30 °C. In part, this is because mixings can be very large and the long processing times required may contribute to excess gassing in the divider, even though divider cutting rates tend to be higher than seen in the UK. Even when CBP mixers were fitted with cooling jackets, acceptance in the USA was limited. Mixing times for US sponge and dough systems are much longer than typically required in the CBP and the extended mixing times seen with the horizontal bar mixers commonly used provided increased time for transfer of heat from dough to cooling jacket. A practical factor which contributed to the non-adoption of the CBP was the smaller size of CBP-type mixer compared with available horizontal-bar-type mixers. This meant that, in order to meet US plant speeds, more than one mixer would be needed for a line and in some bakers' eyes this represented an unwelcome additional expense.

The CBP has been used in many non-English-speaking countries. In many cases its adoption has been limited by the type of bread being produced for the local markets. As commented on previously, many bakers have come to associate (wrongly) the CBP exclusively with the sandwich loaf and so those making other forms of bread, e.g. baguettes, have not seen the CBP as the dough-making process of choice. In the last few years the development of the pressure–vacuum mixer with its greater versatility has generated increased interest in the CBP in many countries. This interest has also increased as the result of greater awareness of the 'British' sandwich and the trend to consumer snacking or 'grazing'. This has even resulted in the introduction of the CBP into France, Germany and Spain.

The range of international products which might use the CBP is wide. It even includes Chinese steamed breads (Cauvain and Huang, 1986). These products are characterised by their round bun-like shape and white-coloured thin delicate crust. The latter arises from the steaming rather than baking of the product in an oven. Cauvain and Huang (1986) showed that the application of the CBP offered a number of advantages over the more traditional fermentation-based processes.

11

The future of
the Chorleywood Bread Process

Bread is a staple foodstuff made and eaten in most countries around the world and, over the many centuries that it has been produced, it has evolved to take many forms. The processes used to make bread have also undergone change and evolved into different forms. Even though it was introduced as recently as 1961, the CBP has also undergone change. Such changes have arisen for many reasons, including the drive for greater efficiency, improved product quality, reduced costs, the arrival of new raw materials and new ingredients, consumer pressures for greater variety and new products, and legislative pressures. However, the main principles of the CBP remain as key to its use as they were when first introduced; they may be summarised as the mixing of the ingredients and development of dough in one single operation such that the dough may be processed to unit-sized dough pieces without the need for a significant resting period before fermentation in the prover.

All breadmaking processes evolve and we can expect that the same will be true for the CBP. Future developments associated with the CBP largely remain a matter of conjecture but there are some obvious and some less obvious opportunities in what to many may be classed as a 'traditional' process. Some speculative thoughts are discussed below.

Raw materials

Critical raw materials for use in the CBP (and indeed any breadmaking process) remain wheat flour and yeast. The quality of the gluten developed in the dough remains crucial in determining dough and final product quality.

Considerable advances have occurred in the development of new wheat varieties for breadmaking and there is no doubt that this will continue to be prominent. However, it has now been fully recognised that wheat varietal development is not simply a matter of pursuing higher protein levels and increasingly stronger gluten. The need for an 'appropriate' wheat and flour protein content will remain an objective for wheat breeders, farmers and flour millers but there will be as great an emphasis on the 'quality' of the protein, particularly in terms of dough rheology and machinability. The key to such developments lies in understanding the genetic basis of wheat varietal performance and it is worth noting that the knowledge links required must take into account the breadmaking process in which the wheat flour will be used. In other words, instead of the current practice of producing a 'breadmaking' wheat variety we may see the exclusive development of new CBP breadmaking varieties.

Continuing studies of the dough development process will no doubt shed new light on the required wheat protein characteristics for the CBP. In this context the hypothesis for the potential formation of 'good' and 'bad' tyrosine cross-links in wheat-flour dough should be considered. The work of Tilley et al. (2001) suggests that the formation of dityrosine is a significant contributor to dough development together with the formation of disulphide and other bonds. At the time of writing, the relative importance of the tyrosine hypothesis to breadmaking in general and to the CBP in particular is uncertain. However, it warrants serious attention, not least because of the potential link with wheat variety and protein quality (in the sense of being able to yield dough and bread of the desired quality).

In the last few years there has been a distinct focus on the genetic basis for dough development with links through to end-use suitability of wheat flour. Many of the more recent studies, especially in the UK, have been based on wheat-flour performance in the CBP. Such more 'generic' approaches will help to avoid some of the 'mistakes' which have occurred in past wheat-breeding programmes, especially those where the emphasis has been on the introduction of 'strong' wheats, many of which have been commercial failures or of limited value to the milling and baking industries. In the UK the wheat variety Hereward enjoys a 'special' status in the minds of miller and bakers; yet the genetic basis for Hereward's popularity is not understood. Research into quality trait loci offers a major step forward in determining the genetic basis for wheat variety performance and should lead to the development of wheat varieties more suited to the needs of bakers using the CBP (Cauvain et al., 2003).

Bakers' yeast has changed little since the introduction of the CBP which necessitated the introduction of new strains to cope with the shorter process times which were used. There have been some developments related to sugar and freeze-tolerant yeast for specific end uses. To some extent it is difficult to foresee major changes in yeast performance. Faster gassing rates would inevitably appeal to those bakers who would like to reduce dough-processing times further but would certainly not appeal to those bakers who associate processing (fermentation) times with flavour development, or lack of it. A fundamental property of yeast is the provision of carbon dioxide gas to inflate the dough; however, if some other means of achieving this inflation were possible, then the need for yeast becomes questionable. Indeed, it may be relegated (or promoted depending on your point of view) to becoming a contributor of flavour. The potential for this changing role for yeast probably lies with modifications to dough mixing and processing (see below).

Functional ingredients

Since its first introduction in 1961 the range of oxidising agents permitted for use in the CBP has decreased in many parts of the world. At the time of writing, the USA retains the greatest list of oxidising agents for potential use in breadmaking; yet the application of the CBP remains limited. As has been referred to on a number of occasions in the text above, the reliance on ascorbic acid in the CBP in the UK and elsewhere has led to modifications of the original format. Enzymes have assumed greater importance in breadmaking in general but especially so in situations where dough-mixing and processing times are relatively short, as is the case with the CBP. Short processing times require enzymes to have far greater specificity than with the bulk fermentation systems in the past. In the CBP, mixing times of 3 or 4 min give little opportunity for significant enzymic activity, even with dough temperatures of 30 °C or so by the end of mixing.

While the potential for developing new enzymes remains significant, the means of doing does not have universal acceptance. Some of the enzymes in current use are derived from micro-organisms which have been genetically modified. Even though no genetically modified material may remain in the purified enzyme product, not all potential enzyme users find this approach acceptable. In the light of the current anxiety regarding the application of genetic modification to human foodstuffs, some enzyme suppliers prefer to scour the world looking for natural sources of new enzymes. Whichever approach finds long-term consumer acceptance, there is no doubt that the identification and use in breadmaking of enzymes with very specific activities will continue to be important for many years. Improved understanding of the functionality of enzymes and their interactions with the various components of wheat flour during the breadmaking process will continue to be a key research theme.

The key to developing new functional ingredients for the CBP probably lies with a better understanding of how the required gluten characteristics and gas bubble population can be created in the mixing cycle. In particular, attention needs to be focused on the mechanism by which gas bubbles become trapped and subsequently stabilised in the dough matrix and the contribution that ingredients make to these properties.

Dough mixing and control

Since doughs are mixed to a fixed work input rather than time in the CBP, there appear to be limited opportunities for further changes to dough mixing and control. Research and practical work over the last few years have shown that the optimum level of work input for a given flour varies. It is common practice for flour millers to grist (i.e. blend together wheats with different milling and functional characteristics to meet a given flour specification) accordingly for CBP flours and this minimises the variability in dough development requirements which may occur as different wheat varieties become available. At present a key method of establishing 'work input requirements' is based on test baking over a range of work inputs and studying the trend in bread quality which occur.

The application of NIR technology offers new opportunities for mixer control (Millar and Alava, 2005). Two possible approaches are proposed. One could be based on the achievement of 'optimum' dough development defined by an NIR parameter measured during the mixing cycle and, once achieved, the mixer could be stopped. To some extent, this is similar to the approach used in the CBP by mixing to a defined energy input independent of time but would have the advantage of adjusting the mixing cycle based on what is happening in the mixer rather than on test baking parameters derived outside the actual production bakery. The definition of a NIR optimum dough development still has to be correlated with some form of 'out-of-plant' test and related to bread quality. Millar and Alava (2005) have shown that this is not as easy as it appears, not least because finest bread cell structure (smallest bread cell size) and maximum bread volume occur at different times in the mixing cycle, the finest cell structure occurring before maximum bread volume is achieved. This may argue a case for a degree of dough 'breakdown' being required in order to achieve maximum bread volume. The NIR dough development optimum proposed by Millar and Alava (2005) occurs roughly midway between the mixing time required for finest cell structure and that for maximum bread volume.

While a number of workers have defined potential NIR dough development optima, a number of practical issues remain to be resolved. The greatest problem is that, in order to obtain a smoothed NIR trace and so to detect the optimum mixing time, the dough needs to be readily visible to the probe for long periods of time. In the Tweedy-type mixers commonly used in the production of dough by the CBP, the bulk of the dough spends longer periods of time being thrown around the bowl and so placing of the NIR probe for signal detection will be critical. The precise means by which this might be achieved have to be defined. Mixing times in the CBP are relatively short, typically 3–5 min, and so data processing must be rapid. The equipment currently used for measurement and data analysis is hardly suitable for introduction into a practical commercial bakery and so significant modification will be required in order to provide a truly commercial application. As will always be the case, bakers will need to be convinced of the real value of expensive measuring equipment which will increase capital costs (cf. the case with watt–hour meters and CBP-type mixers in craft and in-store bakeries discussed above).

The second approach to the application of NIR is that proposed by the American Institute of Baking. They have proposed the application of NIR during mixing for the detection of, or lack of, ingredients during the mixing cycle with the intention of adding in the missing ingredient if possible, or at least being able to reject a dough before further processing. The longer dough mixing times used in the manufacture of US-style bread combined with a mixing action which gives significant opportunity for observing and measuring the dough characteristics would appear to offer an improved potential for the application of NIR to control mixing. The potential for ingredient detection in CBP-style mixers has not been explored.

Dough processing

While the methods used to process dough after mixing are not strictly an essential feature of the CBP, many aspects of the breadmaking process have changed as a consequence of the introduction of the CBP and so it is appropriate to consider what future changes may also occur. A key influence on the way in which dough came to be processed following the introduction of the CBP was the low levels of gas that were present in the dough reaching the divider by comparison with the then common bulk fermented doughs. Thus degassing of CBP doughs was never necessary. It has taken some while for equipment companies and bakers to understand that denser CBP doughs required less pressure during dividing in order to achieve consistent divider performance. Some of the more recently introduced dough dividers put less pressure on the dough (sometimes called 'stress-free' dividers but this is a misnomer since any mechanical action on bread dough will put it under some degree of stress) and avoid the inadvertent 'damage' to the microscopic dough structures. Even though CBP doughs are more dense than many doughs, the gluten structure at the microscopic level remains relatively delicate and can be broken down during mechanical processing.

A number of workers (Cauvain and Collins, 1995; Cauvain and Little, 2001) have questioned the value of the dough-processing steps between dividing and end of final moulding. Aspects of this work have been discussed in Chapter 6. An increasing number of CBP dough-processing plants have been modified to eliminate first moulding and to reduce intermediate proof without the loss of product quality. However, the ability to do this does depend on the interactions among ingredient qualities, dough formulation and processing conditions; the use of CBP alone does not mean that such dramatic changes can be made to dough processing.

The identification of the potential for the final moulder to contribute to the loss of bread quality has caused greater attention to be given to final moulder design. Once again any future developments will rely on a complete understanding of all aspects of dough production in order to develop more truly 'dough-friendly' final moulders.

Proving and baking

By the time that a CBP dough reaches the final prover and then the oven a great many of the essential quality attributes will be in place. CBP doughs behave in a very similar manner to all other fermented doughs, namely they will expand in proof, expand again in the oven and then be heat set. Further changes in the design and operation of provers and ovens may well occur in the future but they are unlikely to be as a direct consequence of the use of the CBP. The main challenge which faces the development of improvements to proving and baking operations is the relatively poor conductivity of bread dough. The larger the dough piece (e.g. larger diameter), the greater is the challenge.

There have been a number of documented improvements to prover and oven design (Wiggins, 1998) and these have often been associated with plant bakeries using the CBP. Microwave-assisted baking has been investigated in the past (Chamberlain, 1973) with some success. Chamberlain showed that one potential benefit of using microwave-assisted baking was that the flour Falling Number became a less critical parameter for the manufacture of sliced sandwich bread. In the past the uptake of microwave-assisted baking was limited by the lack of availability of suitable non-metallic pans; technology has moved a long way forward and so microwave-assisted baking may yet become a commercial reality.

12

Conclusions

Over the centuries, craft bakers around the world have developed our traditional bread varieties using their accumulated knowledge as to how to make best use of their available raw materials in order to achieve given bread qualities. In some countries the nature of breadmaking has retained its traditional forms while in others it has changed dramatically. The introduction of the CBP had a profound effect on the direction in which breadmaking processes and baking technology progressed in the UK and elsewhere. Before the 1960s, no-time dough-making processes were known to bakers although seldom used in the normal course of events in most bakeries. Recipes for 'emergency' no-time doughs were available for bakers who might find that suddenly they had to produce extra bread or who were having problems with normal process control techniques. However, such recipes were only expected to be used in an 'emergency' and the bread that it yielded was expected to have inferior quality, with lower volume, poorer shape, firmer crumb and more rapid staling properties.

As is well known, the character of bread and other fermented products depends heavily on the formation of a gluten network in the dough. In bulk fermentation processes, time, temperature, yeast level and flour quality have to be optimised in order to achieve the required quality and production consistency. Emergency no-time dough processes could not deliver the required product quality because the 'time' effects were no longer applicable. In the CBP, the replacement of the fermentation time effect with energy during mixing represented a significant step forwards in achieving a more consistent dough and in turn more consistent bread quality. Consistency of bread quality was and still is a major factor which influences consumers' choice of which bread product they will buy, although the precise definition of the 'right' quality will remain highly individual.

To be able to make our particular bread type we must have an understanding of the complex interactions between the raw materials and methods which we shall use in the conversion processes involved in the move from ingredients to final baked product. This is true whether one is using the CBP or any other breadmaking process that has been evolved. The 'improvement' in bread volume obtained for a given flour when moving from a bulk fermentation process to the CBP offered bakers an opportunity to adjust flour properties in the complex relationship that is breadmaking. Unfortunately the opportunities which allowed a reduction in flour protein content have become, in some eyes, synonymous with all that is 'wrong' with the CBP. It is ironic that the same critics have overlooked the fact that craft bakers of the 1950s would regularly adjust the length of the bulk fermentation time according to the 'strength' of the flour. It has long been recognised that weaker flours required shorter fermentation times to achieve optimum bread quality and were less tolerant to processing delays. In the past, emergency no-time doughs were expected to be made with weaker flours than 'normal' and, if not, then it was expected that stronger flours would not yield optimum bread quality with reduced fermentation times.

So why, if the principles of matching flour quality with variable dough development times were known, was the CBP so denigrated by craft bakers when first introduced to the UK baking industry and why does it still suffer from similar criticisms today? The answer is clear: because many bakers do not understand the fundamentals which underpin successful bread production and so, when a new concept such as the CBP arrived in the marketplace, it was (and still is today) treated with scepticism and fear – fear because many bakers saw and indeed still consider that the introduction of a breadmaking process based on scientific understanding will demystify the art that craft bakers have practised for many years. It may be an overstatement but the introduction of the CBP unleashed the Luddites of the baking industry. We would suggest that such attitudes have more to do with the demise of the craft baker in the UK than the introduction of the 'infamous' CBP.

One example which supports this view when the CBP was first launched was the attitude towards bread improvers. Typically the value of increased oxidation in the CBP was greeted with the usual antagonism that goes with the use of 'additives'. Yet the same anti-CBP lobby consisted of bakers using flour treated with potassium bromate, and in many cases chlorine dioxide gas, in the flour mill, the most common forms at the time of adjusting the bread-making properties of flours being used in bulk fermentation systems. The passage of time has reduced the list of permitted improvers for breadmaking in the UK and both potassium bromate and chlorine dioxide have long been abandoned in the UK baking industry. Ascorbic acid (vitamin C), ironically the improver on which the CBP was launched, remains in common use by craft and plant bakers alike. However, in the less well-informed popular press it is its use in the CBP that is the subject of attack.

It is inevitable that the raw materials and processes used by bakers will change with time. Much of the change is being driven by consumers of baked products. If it were not so, then we would all still be eating the unleavened barley and wild-grass breads of some 12 000 years ago. Bakers have learnt how to adapt and change their products to stay in business. The introduction of the CBP was an example of the inventiveness of bakers as they responded to consumer demands for bread of consistent quality with improved keeping properties and a reasonable price. Bread made by the CBP is just as much 'bread' as that made by any other breadmaking process and is another illustration of the wide-ranging variety of products that can be made from the versatile cereal that we call wheat and the adaptability of bakers.

References

Alava, J.M., Millar, S.J. and Salmon, S.E. (2001) The determination of wheat bread-making performance and dough mixing time by NIR spectroscopy for high speed mixers. *Journal of Cereal Science*, **33**, 71–81.

Anon. (1983) High-speed, high-energy mixing units make inroads on conventional systems. *Milling and Baking News*, February, BE-20, BE-22.

Anon. (1988) Baking progress. *Food Processing*, April, 37–38.

APV Corporation Ltd (1992) Dough mixing. *UK Patent GB 2,264,623A*, HMSO, London.

Baker, J.C. and Mize, M.D. (1941) The origin of the gas cell in bread dough. *Cereal Chemistry*, **18**, 19–34.

Baker, J.C. and Mize, M.D. (1942) The relationship of fats, texture, crumb and volume in bread. *Cereal Chemistry*, **19**, 84–94.

Bent, A. (1998) Speciality fermented goods. In S.P. Cauvain and L.S. Young (Eds), *Technology of Breadmaking*, Blackie, London, pp. 214–239.

Brooker, B.E. (1996) The role of fat in the stabilisation of gas cells in bread dough. *Journal of Cereal Science*, **24**, 187–198.

Brown, J. (1993) Advances in breadmaking technology. In B.S. Kamel and C.E. Stauffer (Eds), *Advances in Baking Technology*, Blackie, Glasgow, pp. 38–87.

Campbell, G.M. (2003) Bread aeration. In S.P. Cauvain (Ed.), *Bread Making: Improving Quality*, Woodhead Publishing, Cambridge pp. 352–374.

Cauvain, S.P. (1981) The production, use and practical application of frozen bread dough. *FMBRA Report 94*, CCFRA, Chipping Campden, Gloucestershire.

Cauvain, S.P. (1985) Effects of some enzymes on loaf volume in the CBP. *FMBRA Bulletin 1*, CCFRA, Chipping Campden, Gloucestershire, pp. 11–17.

Cauvain, S.P. (1987) Effects of bran, germ and low grade flours on CBP bread quality. *FMBRA Report 138*, CCFRA, Chipping Campden, Gloucestershire.

Cauvain, S.P. (1994) New mixer for variety bread production. *The European Food and Drink Review*, Autumn, 51, 53.

Cauvain, S.P. (1995) Creating the structure: the key to quality. *South African Food Review*, **22**, 33, 35, 37.

Cauvain, S.P. (1997) Controlling the structure is the key to quality. In *Proceedings of the Fiftieth Anniversary Meeting of the Australian Society of Baking*, Sydney, Australia, 16 October 1997, Australian Society of Baking, Sydney, pp. 6–11.

Cauvain, S.P. (1998a) Breadmaking processes. In S.P. Cauvain and L.S. Young (Eds), *Technology of Breadmaking*, Blackie, London, pp. 18–44.

Cauvain, S.P. (1998b) Bread – the product. In S.P. Cauvain and L.S. Young (Eds), *Technology of Breadmaking*, Blackie, London, pp. 1–17.

Cauvain, S.P. (1998c) Dough retarding and freezing. In S.P. Cauvain and L.S. Young (Eds), *Technology of Breadmaking*, Blackie, London, pp. 149–178.

Cauvain, S.P. (2001) Breadmaking. In G. Owens (Ed.), *Gereals Processing Technology*, Woodhead Publishing, Cambridge, pp. 204–230.

Cauvain, S.P. and Chamberlain, N. (1988) The bread improving effect of fungal alpha-amylase. *Journal of Cereal Science*, **8**, 239–248.

Cauvain, S.P and Collins, T.H. (1995) Mixing, moulding and processing bread doughs. In A. Gordon (Ed.), *Baking Industry Europe*, Sterling Publications, London, pp. 41–43.

Cauvain, S.P., Collins, T.H. and Pateras, I. (1992) Effects of ascorbic acid during dough processing. *Chorleywood Digest 121*, CCFRA, Chipping Campden, Gloucestershire, pp. 111–114.

Cauvain, S.P, Davies, J.A. and Fearn, T. (1985) Flour characteristics and fungal alpha-amylase in the Chorleywood Bread Process. *FMBRA Report 121*, CCFRA, Chipping Campden, Gloucestershire.

Cauvain, S.P and Huang, S. (1986) Chinese steamed bread. *FMBRA Bulletin 4*, CCFRA, Chipping Campden, Gloucestershire, pp. 151–158.

Cauvain, S.P. Law, C., Bhandari, D., Salman, S. and Worland, A.J. (2003) Characterisation of a novel genetic contribution to the breadmaking quality of wheat. *HGCA Project Report 276*, HGCA, London.

Cauvain, S.P. and Little, K. (1999) Mixing and processing Chorleywood Bread Process doughs: 1. Factors affecting dough rheology. *CCFRA R&D Report 115*, CCFRA, Chipping Campden, Gloucestershire.

Cauvain, S.P. and Little, K. (2000) Mixing and processing Chorleywood Bread Process doughs: 2. The application of pressure–vacuum mixers. *CCFRA R&D Report 115*, CCFRA, Chipping Campden, Gloucestershire.

Cauvain, S.P. and Little, K. (2001) Mixing and processing Chorleywood Bread Process doughs: 3. Influences of processing stages on bread quality. *CCFRA R&D Report 120*, CCFRA, Chipping Campden, Gloucestershire.

Cauvain, S.P., Whitworth, M.B. and Alava, J.M. (1999) The evolution of bubble structure in bread doughs and its effects on bread cell structure. In G.M. Campbell, C. Webb, S.S. Pandiella and K. Niranjan (Eds), *Bubbles in Food*, Eagen Press, St Paul, Minnesota, pp. 85–88.

Cauvain, S.P. and Young, L.S. (1998) *Technology of Breadmaking*, Blackie, London.

Cauvain, S.P. and Young, L.S. (2000) *Bakery Food Manufacture and Quality: Water Control and Effects*, Blackwell Science, Oxford.

Cauvain, S.P. and Young, L.S. (2001) *Baking Problems Solved*, Woodhead Publishing, Cambridge.

Chamberlain, N. (1973) Microwave energy in the baking of bread. *FMBRA Bulletin 3*, CCFRA, Chipping Campden, Gloucestershire, pp. 76–87.

Chamberlain, N. (1979) Gases – the neglected bread ingredients. In *Proceedings of the 49th Conference of the British Society of Baking*, London, 1979, British Society of Baking, London, pp. 12–17.

Chamberlain, N. (1983) U.K. researcher describes evolution of Chorleywood Process, Tweedy mixer. *Milling and Baking News*, **61**, BE-26, BE-28, BE-30.

Chamberlain, N. (1984) Dried gluten in breadmaking – the new challenge. *British Society of Breadmaking, 30th Annual Meeting and 59th Conference*

Proceedings, November 1984, British Society of Breadmaking, Birmingham, pp. 14–18.

Chamberlain, N. (1985) Dough formation and development. In J. Brown (Ed.), *The Master Bakers Book of Breadmaking*, 2nd edition, Turret–Wheatland, Rickmansworth, Hertfordshire, pp. 47–57.

Chamberlain, N. and Collins, T.H. (1979) The Chorleywood Bread Process: the roles of oxygen and nitrogen. *Bakers Digest*, **53**, 18–24.

Chamberlain, N. and Collins, T.H. (1981) The Chorleywood Bread Process: oxidizing improver effects of potassium bromate and ascorbic acid. *FMBRA Report 95*, CCFRA, Chipping Campden, Gloucestershire.

Chamberlain, N. and Collins, T.H. (1984) The Chorleywood Bread Process: replacement of bromate with ascorbic acid in commercial trials. *PMBRA Report 119*, CCFRA, Chipping Campden, Gloucestershire.

Chamberlain, N., Collins, T.H. and Elton, G.A.H. (1961) The Chorleywood Bread Process. *BBIRA Report 59*, CCFRA, Chipping Campden, Gloucestershire.

Chamberlain, N., Collins, T.H. and Elton, G.A.H. (1965) The Chorleywood Bread Process: the improving effects of fat. *Cereal Science Today*, **10**, 415–416, 418–419, 490.

Chamberlain, N., Collins, T.H. and Elton, G.A.H. (1967) The Chorleywood Bread Process: effect of rate of dough mixing. *FMBRA Report 1*, CCFRA, Chipping Campden, Gloucestershire.

Chamberlain, N., Collins, T.H. and Mcdermott, E.E. (1977) The Chorleywood Bread Process: the effects of alpha-amylase activity on commercial bread. *FMBRA Report 73*, CCFRA, Chipping Campden, Gloucestershire.

Collins, T.H. (1964) The Chorleywood Bread Process: variety productions. *BBIRA Report 74*, CCFRA, Chipping Campden, Gloucestershire.

Collins, T.H. (1978) Making French bread by the CBP. *FMBRA Bulletin 6*, CCFRA, Chipping Campden, Gloucestershire, pp. 193–201.

Collins, T.H. (1986) Oxygen and the Chorleywood Bread Process. *FMBRA Bulletin 6*, CCFRA, Chipping Campden, Gloucestershire, pp. 239–250.

Collins, T.H. (1993) The Chorleywood Bread Process: pressure/vacuum mixing. *Chorleywood Digest 130*, CCFRA, Chipping Campden, Gloucestershire, pp. 94–97.

Collins, T.H., Devlin, J.J. and James, A.W. (1959) The use of brews in continuous and conventional mixing processes for bread doughs. *BBIRA Report 42*, CCFRA, Chipping Campden, Gloucestershire.

Coins, T.H., Devlin, J.J. and James, A.W. (1960) A study of the factors affecting the production of bread from mechanically developed dough. *BBIRA Report 47*, CCFRA, Chipping Campden, Gloucestershire.

Collins, T.H. and Evans, K. (1982) Chorleywood Bread Process: effects of fungal alpha-amylase and malt flour. *FMBRA Bulletin 2*, CCFRA, Chipping Campden, Gloucestershire, pp. 55–60.

Collins, T.H. and Evans, K. (1984) Chorleywood Bread Process: loaf volume improvement from gluten addition to flour. *FMBRA Bulletin 2*, CCFRA, Chipping Campden, Gloucestershire, pp. 43–47.

Collins, T.H. and Fearn, T. (1986) Comparison of gluten and Canadian wheat flour in white bread made using the Chorleywood Bread Process. *FMBRA Bulletin 2*, CCFRA, Chipping Campden, Gloucestershire, pp. 56–62.

Collins, T.H., Fearn, T. and Ford, W. (1985) The effects of gluten, fungal alpha-amylase and DATA ester in wholemeal bread made by CBP. *FMBRA Bulletin 5*, CCFRA, Chipping Campden, Gloucestershire, pp. 194–201.

Collins, T.H. and Ford, W. (1985) Part-baked soft rolls and French bread. *FMBRA Bulletin 3*, CCFRA, Chipping Campden, Gloucestershire, pp. 128–134.

Collins, T.H. and Haley, S. (1992) Frozen bread doughs: effect of ascorbic acid addition and dough mixing temperature on loaf properties. *Chorleywood Digest 114*, CCFRA, Chipping Campden, Gloucestershire, pp. 21–23.

Collins, T.H., Little, K and Pritchard, P.E. (1991) Dehydroascorbic acid; potential oxidising improver in the CBP? *FMBRA Bulletin 4*, CCFRA, Chipping Campden, Gloucestershire, pp. 91–98.

Collins, T.H., Overton, M.J., Redman, B.I., Roberts, J.A. and Smith, M.J. (1968) A summary of breadmaking processes. *FMBRA Report 13*, CCFRA, Chipping Campden, Gloucestershire.

Collins, T.H., Redman, B.I. and Turner, M.L. (1971) Part-baked bread. *FMBRA Report 48*, CCFRA, Chipping Campden, Gloucestershire.

Collins, T.H. and Sang, B. (1993) Loaf side crust caving or collapse. *Chorleywood Digest 126*, CCFRA, Chipping Campden, Gloucestershire, pp. 49–50.

Collins, T.H. and Smith, M.J. (1976) The use of soya flour in the Chorleywood Bread Process. *FMBRA Bulletin 2*, CCFRA, Chipping Campden, Gloucestershire, pp. 58–60.

Collins, T.H. and Tomlinson, D. (1983) Dough mixing for farinaceous foodstuff production. *US Patent 4,369,193*.

Collins, T.H. and Young, L.S. (1986) Gluten fortification of brown flours used in the Chorleywood Bread Process. *FMBRA Bulletin 3*, CCFRA, Chipping Campden, Gloucestershire, pp. 95–101.

Collyer, D.M. (1967) Sugar brews and the Chorleywood Bread Process. *Journal of Science, Food and Agriculture*, **18**, 596–601.

Connolly, P. and Pittman, M. (1997) *Practical Bakery*, Hodder & Stoughton, London.

David, E. (1977) *English Bread and Yeast Cookery*, Penguin Books, London.

Davidson, S. (1995) *Loaf, Crust and Crumb*, Michael Joseph, London.

Dodds, N.J.H. and Collins, T.H. (1971) The Chorleywood Bread Process; some effects of malt flour and fungal alpha-amylase. *FMBRA Report 50*, CCFRA, Chipping Campden, Gloucestershire.

Elton, G. and Devlin, J. (1959) Mechanical dough development processes for large and small bakers. *Bakers' Review*, **25**, 1047–1049.

Evans, K. (1984) CBP: effect of emulsifiers on the quality of soft rolls. *FMBRA Bulletin 4*, CCFRA, Chipping Campden, Gloucestershire, pp. 160–163.

Faridi, H. and Faubion, J.M. (1989) *Dough Rheology and Baked Product Texture*, Van Nostrand Reinhold – AVI, New York.

Ford, W.P. (1975) Earlier dough development processes. In A. Williams (Ed.), *Breadmaking: The Modern Revolution*, Hutchinson Beenham, London, pp. 13–24.

French, F.D. and Fisher, A.R. (1981) High speed mechanical dough development. *Bakers Digest*, **55**, 80–82.

Gould, J.T. (1998) Breadmaking around the world. In S.P. Cauvain and L.S. Young (Eds), *Technology of Breadmaking*, Blackie, London, pp. 197–213.

HMSO (1988) *The Weights and Measures (Miscellaneous Foods) Order 1998* (SI 1988, No. 2040), HMSO, London.

Kamel, B.S. and Ponte, J.G. (1993) Emulsifiers in baking. In B.S. Kamel and C.E. Stauffer (Eds), *Advances in Baking Technology*, Blackie, London, pp. 179–222.

Kilborn, R.H. and Tipples, K.H. (1974) Implications of the mechanical development of bread dough by means of sheeting rolls. *Cereal Chemistry*, **51**, 648–657.

Kulp, K. (1993) Enzymes as dough improvers. In B.S. Kamel and C.E. Stauffer (Eds), *Advances in Baking Technology*, Blackie, London, pp. 152–178.

Lawrence, F. (2004) *Not on the Label: What Really Goes into the Food on your Plate*, Penguin Books, London.

Marsh, D. (1998) Mixing and dough processing. In S.P. Cauvain and L.S. Young (Eds), *Technology of Breadmaking*, Blackie, London, pp. 81–119.

Melville, J. and Shattock, H.T. (1938) The action of ascorbic acid as a bread improver. *Cereal Chemistry*, **15**, 201.

Millar, S. (2003) Controlling dough development. In S.P. Cauvain (Ed.), *Bread Making: Improving Quality*, Woodhead Publishing, Cambridge, pp. 401–423.

Millar, S.J. and Alava, J.M. (2005) Flow quality and dough development interactions – the critical first steps in bread production. In S.P. Cauvain, S.E. Salmon and L.S. Young (Eds), *Using Cereal Science and Technology for the Benefit of Consumers*, Woodhead Publishing, Cambridge, pp. 132–136.

Pateras, I. (1998) Bread spoilage and staling. In S.P. Cauvain and L.S. Young (Eds), *Technology of Breadmaking*, Blackie, London, pp. 240–261.

Petryszak, R., Young, L. and Cauvain, S. (1995) Improving cake product quality. In *Proceedings of Expert Systems 95, the 15th Annual Technical Conference of the British Computer Society Specialist Group on Expert Systems*, December 1995, SGES Publications, Oxford, pp. 161–168.

Pickles, K. (1968) Tweedy (Chipping) Ltd, Improvements in or relating to dough production. *UK Patent 1,133,472*, HMSO, London.

Russell Eggitt, P.W. (1975) Dough consistency control. *Bakers Digest*, June, 31–35.

Stauffer, C.E. (1999) *Fats and Oils*, Eagan Press, St Paul, Minnesota.

Stolz, P. and Bocker, G. (1996) Technology, properties and applications of sourdough products. *Advances in Food Science*, **18**, 234–236.

Swanson, C.O. and Working, E.B. (1926) Mechanical development of dough. *Cereal Chemistry*, **3**, 65.

Thompson, D.R. (1983) Liquid sponge technology applied to high-speed dough mixing. *Bakers Digest*, November, 11, 12, 14, 16, 17.

Tilley, K.A., Benjamin, R.E., Bagoragoza, M., Okot-Kotber, B., Prakash, O. and Kwen, H. (2001) Tyrosine crosslinks: molecular basis of gluten structure and function. *Journal of Agricultural and Food Chemistry*, **49**, 2627–2632.

Tiphaine, L., Lucas, T., Grenier, A., Quellec, S., Collewet, G., Davenel, A. and Le Bail, A. (2005) Use of MRI for the characterization of the bread process. In S.P. Cauvain, S.S. Salmon and L.S. Young (Eds), *Using Cereal Science and Technology for the Benefit of Consumers*, Woodhead Publishing, Cambridge, pp. 199–204.

Tomlinson, D. (1983) Dough mixing for farinaceous foodstuffs. *US Patent 4,368,209*.

Tweedy of Burnley Ltd (1982) Dough mixing for farinaceous foodstuffs. *UK Patent GB 2,030,883B*, HMSO, London.

Wheelock, T.D. and Lancaster, E.B. (1970) Thermal properties of wheat flour. *Starke*, **22**, 44–48.

Whitworth, M.B. and Alava, J.M. (1998) Determination of critical bubble size in bread doughs. *CCFRA R&D Report 65*, CCFRA, Chipping Campden, Gloucestershire.

Whitworth, M.B. and Alava, J.M. (1999) The imaging and measurement of bubbles in bread doughs. In G.M. Campbell, C. Webb, S.S. Pandiella and K. Niranjan (Eds), *Bubbles in Food*, Eagen Press, St Paul, Minnesota, pp. 221–231.

Whitworth, M. and Alava, J. (2002) Factors affecting the stability of gas bubbles in dough. *CCFRA R&D Report 155*, CCFRA, Chipping Campden, Gloucestershire.

Whitworth, M., Cauvain, S. and Cliffe, D. (2005) Measurement of bread cell structure by image analysis. In S.P. Cauvain, S.S. Salmon and L.S. Young (Eds), *Using Cereal Science and Technology for the Benefit of Consumers*, Woodhead Publishing, Cambridge, pp. 193–198.

Wiggins, C. (1998) Proving, baking and cooling. In S.P. Cauvain and L.S. Young (Eds), *Technology of Breadmaking*, Blackie, London, Chapter 5, pp. 120–148.

Williams, A. (1975) *Bread Making: The Modern Revolution*. Hutchinson Benham, London.

Williams, A. and Pullen, G. (1998) Functional ingredients. In S.P. Cauvain and L.S. Young (Eds), *Technology of Breadmaking*, Blackie, London, pp. 45–80.

Wilson, A., Morgenstern, M., Pfahringer, B. and Leschi, C. (2005) Data mining bread quality and process data in a plant bakery. In S.P. Cauvain, S.S. Salmon and L.S. Young (Eds), *Using Cereal Science and Technology for the Benefit of Consumers*, Woodhead Publishing, Cambridge, pp. 383–388.

Young, L.S. (1991) The FMBRA Bread Faults Expert System. *Manufacturing Intelligence* 6, Spring, Department of Trade and Industry, London, pp.18–19.

Young, L. and Cauvain, S. (1994) Advising the baker. In *Proceedings of Expert Systems 94, the 14th Annual Conference of the British Computer Society Specialist Group on Expert Systems*, December 1994, SGES Publications, Oxford, pp. 21–33.

Young, L. and Cauvain, S. (2001) Arise again fair knowledge. In *Proceedings of Expert Systems 2001, the 21st SGES International Conference on Knowledge Based Systems and Artificial Intelligence*, December 2001, Springer, London, pp. 89–99.

Young, L., Davies, P.R. and Cauvain, S. (1998) Cakes – getting the right balance. In *Proceedings of Expert Systems 1998, the 18th Annual International Conference of the British Computer Society Specialist Group on Expert Systems*, December 1998, Springer, Berlin, pp. 42–55.

Further reading

Much has been written about the CBP since its invention. The original material is contained in a number of research reports produced by BBIRA, FMBRA and CCFRA. In chronological order they are as follows:

Collins, T.H., Devlin, J.J. and James, A.W. (1959) The use of brews in continuous and conventional mixing processes for bread doughs. *BBIRA Report 42*, CCFRA, Chipping Campden, Gloucestershire.

Collins, T.H., Devlin, J.J. and James, A.W. (1960) A study of the factors affecting the production of bread from mechanically developed dough. *BBIRA Report 47*, CCFRA, Chipping Campden, Gloucestershire.

Chamberlain, N. and Collins, T.H. (1960) The mechanical development of bread doughs: comparison between batch and continuous development on the laboratory scale. *BBIRA Report 48*, CCFRA, Chipping Campden, Gloucestershire.

Collins, T.H. (1960) The mechanical development of bread doughs: effects of dough water content. *BBIRA Report 55*, CCFRA, Chipping Campden, Gloucestershire.

Collins, T.H. (1961) The mechanical development of bread doughs: effect of flour quality. *BBIRA Report 56*, CCFRA, Chipping Campden, Gloucestershire.

Bennett, R. (1961) The mechanical development of bread doughs: laboratory measurement of flour properties. *BBIRA Report 57*, CCFRA, Chipping Campden, Gloucestershire.

Chamberlain, N., Collins, T.H. and Elton, G.A.H. (1961) The Chorleywood Bread Process. *BBIRA Report 59*, CCFRA, Chipping Campden, Gloucestershire.

Chamberlain, N., Collins, T.H., Elton, G.A.H. and Cornford, S.J. (1961) The Chorleywood Bread Process: commercial application. *BBIRA Report 62*, CCFRA, Chipping Campden, Gloucestershire.

Chamberlain, N., Collins, T.H. and Elton, G.A.H. (1963) The Chorleywood Bread Process: choice of flour. *BBIRA Report 66*, CCFRA, Chipping Campden, Gloucestershire.

Collins, T.H. (1964) The Chorleywood Bread Process: variety production. *BBIRA Report 74*, CCFRA, Chipping Campden, Gloucestershire.

Chamberlain, N., Collins, T.H. and Elton, G.A.H. (1965) The Chorleywood Bread

Process: the effect of flour strength on the quality of plant bread. *BBIRA Report 77*, CCFRA, Chipping Campden, Gloucestershire.

Chamberlain, N., Collins, T.H. and Elton, G.A.H. (1965) The Chorleywood Bread Process: the effect of flour strength compared with the bulk fermentation process. *BBIRA Report 82*, CCFRA, Chipping Campden, Gloucestershire.

Chamberlain, N., Collins, T.H. and Elton, G.A.H. (1965) The Chorleywood Bread Process: choice of fat. *BBIRA Report 85*, CCFRA, Chipping Campden, Gloucestershire.

Chamberlain, N., Collins, T.H. and Elton, G.A.H. (1967) The Chorleywood Bread Process: effect of rate of dough mixing. *FMBRA Report 1*, CCFRA, Chipping Campden, Gloucestershire.

Chamberlain, N., Collins, T.H. and Redman, B.I. (1970) The Chorleywood Bread Process: effect of reduced pressure during mixing. *FMBRA Report 40*, CCFRA, Chipping Campden, Gloucestershire.

Dodds, N.J.H. and Collins, T.H. (1971) The Chorleywood Bread Process: some effects of malt flour and fungal alpha-amylase. *FMBRA Report 50*, CCFRA, Chipping Campden, Gloucestershire.

Collins, T.H. (1976) The Chorleywood Bread Process: variety production. *FMBRA Report 71*, CCFRA, Chipping Campden, Gloucestershire.

Chamberlain, N., Collins, T.H. and Mcdermott, E.E. (1977) The Chorleywood Bread Process: the effects of alpha-amylase activity on commercial bread. *FMBRA Report 73*, CCFRA, Chipping Campden, Gloucestershire.

Chamberlain, N. and Collins, T.H. (1981) The Chorleywood Bread Process: oxidising improver effects of potassium bromate and ascorbic acid. *FMBRA Report 95*, CCFRA, Chipping Campden, Gloucestershire.

Chamberlain, N. and Collins, T.H. (1984) The Chorleywood Bread Process: replacement of bromate with ascorbic acid in commercial trials. *FMBRA Report 119*, CCFRA, Chipping Campden, Gloucestershire.

Cauvain, S.P., Davies, J.A. and Fearn, T. (1985) Flour characteristics and fungal alpha-amylase in the Chorleywood Bread Process. *FMBRA Report 121*, CCFRA, Chipping Campden, Gloucestershire.

Collins, T.H. and Davies, J.A. (1985) The Chorleywood Bread Process: control of loaf volume. *FMBRA Report 125*, CCFRA, Chipping Campden, Gloucestershire.

Cauvain, S.P. and Mitchell, T.J. (1986) Effects of gluten and fungal alpha-amylase on CBP bread crumb properties. *FMBRA Report 134*, CCFRA, Chipping Campden, Gloucestershire.

Cauvain, S.P. (1987) Effects of bran, germ and low grade flour on CBP bread quality. *FMBRA Report 138*, CCFRA, Chipping Campden, Gloucestershire.

Cauvain, S.P. and Little, K. (1999) Mixing and Processing Chorleywood Bread Process doughs: 1. Factor affecting dough rheology. *CCFRA R&D Report 84*, CCFRA, Chipping Campden, Gloucestershire.

Cauvain, S.P., Little, K. and Anfield, J. (2001) Mixing and processing Chorleywood Bread Process doughs: 2. The application of pressure–vacuum mixers. *CCFRA R&D Report 115*, CCFRA, Chipping Campden, Gloucestershire.

Cauvain, S.P. and Little, K. (2001) Mixing and processing Chorleywood Bread Process doughs: 3. Influences of processing stages on bread quality. *CCFRA R&D Report 120*, CCFRA, Chipping Campden, Gloucestershire.

Some of the more contemporary accounts of the development of the CBP and related issues can be read in the following:

Anon. (1972) British Baking Industries Research Association: Chorleywood Bread Process. In J. Langrish, M. Gibbons, W.G. Evans and F.R. Jevons (Eds), *Wealth from Knowledge: A study of Innovation in Industry*, Macmillan, London, pp. 136–149.

The article provides the background to some of the clues and the thinking that were to lead to the development of the CBP. To quote, 'This study is an illustration of how old ideas can be made the basis of an important new method of manufacture.'

Williams, A. (1975) *Breadmaking: The Modern Revolution*, Hutchinson Benham, London.
Out of print. This provides an early and comprehensive review of the development of the CBP and its applications.

Spicer, A. (1975) *Bread: Social, Nutritional and Agricultural Aspects of Wheaten Bread*, Applied Science, London.
This is the proceedings of an international symposium organized by Rank Prize Funds.

Chamberlain, N. (1983) U.K. researcher describes evolution of Chorleywood Process, Tweedy mixer. *Milling and Baking News*, **61**, BE-26, BE-28, BE-30.
This gives a résumé of the key events in the development of the CBP and the Tweedy mixer that became so closely associated with it.

Brown, J. (1985) Chorleywood and the compound revolution. *Baking Today*, April, 6–9.
This reviews the changing role of improvers and their formulation following the introduction of the CBP.

Other relevant publications

Other relevant publications are as follows, in chronological order:

Collins, T.H. (1983) The creation and control of bread crumb cell structure. *FMBRA Report 104*, CCFRA, Chipping Campden, Gloucestershire.

Kamel, B.S. and Stauffer, C.E. (1993) *Advances in Baking Technology*, Blackie, London.

Cauvain, S.P. and Young, L.S. (1998) *Technology of Breadmaking*, Blackie, London.

Cauvain, S.P. and Young, L.S. (1998) *Fabricacion de Pain*, Kluwer–Plenum, New York.

Cauvain, S. and Young, L. (2000) *Bakery Food Manufacture and Quality: Water Control and Effects*, Blackwell Science, Oxford.

Cauvain, S.P. (2003) *Bread making: Improving Quality*, Woodhead Publishing, Cambridge.

Index